MONKEY TRIALS AND GORILLA SERMONS

NEW HISTORIES OF SCIENCE, TECHNOLOGY,
AND MEDICINE

SERIES EDITORS

Margaret C. Jacob and Spencer R. Weart

PETER J. BOWLER

MONKEY TRIALS AND GORILLA SERMONS

EVOLUTION AND CHRISTIANITY FROM DARWIN TO INTELLIGENT DESIGN

HARVARD UNIVERSITY PRESS

CAMBRIDGE, MASSACHUSETTS

LONDON, ENGLAND

2007

Library of Congress Cataloging-in-Publication Data

Bowler, Peter J.
Monkey trials and gorilla sermons : evolution and Christianity from Darwin
to intelligent design / Peter J. Bowler.
p.   cm. — (New histories of science, technology, and medicine)
Includes bibliographical references and index.
ISBN-13: 978-0-674-02615-5 (alk. paper)
ISBN-10: 0-674-02615-2 (alk. paper)
1. Creationism. 2. Evolution (Biology)—Religious aspects—Christianity.
I. Title.   BS651.B755 2007
231.7′652—dc22   2007011200

CONTENTS

My purpose in writing a book on such a controversial topic is explained in the first chapter and need not be repeated here. Suffice it to say that I have been working on the history of Darwinism and its implications for most of my career, and I hope that a balanced historical account of the debate over the theory's religious implications will be of some interest to those engaged in the current controversies. Perhaps by writing from a perspective shaped by residence outside North America, I can shed some light on the less confrontational aspects of the interaction.

Most of what follows is a distillation of my own and other scholars' work over the past years. If there is anything original here, it is my brief foray into the work of the early twentieth-century Modernists within the American churches, whose writings are often neglected in accounts of the period defined in the popular imagination by the Scopes "Monkey Trial."

For their help and inspiration I am particularly indebted to John Brooke, Jim Moore, Ron Numbers, Ted Davis, and Michael Ruse. I am also grateful to Spencer Weart and two anonymous referees for their comments on the original manuscript.

MONKEY TRIALS AND GORILLA SERMONS

THE MYTHS OF HISTORY

There is a widespread assumption that science and religion are at war with one another. Which side deserves to win, of course, depends on your point of view. For many traditional religious believers, science is the agent of Godless materialism hell-bent (literally) on destroying humanity's faith in its Creator. For the humanist or atheist, science is a weapon in the fight to replace ancient superstitions with a rational analysis of our place in the universe. The trial of Galileo by the Inquisition of the Roman Catholic Church is often seen as the opening battle in this war, especially by the humanists, who point out that here even the Church eventually had to agree that its attempt to protect the traditional worldview was misguided. The earth really does go around the sun, whatever may be implied by passages in the Bible (e.g., Joshua 10:13). But the clash of ideas and ideologies centered on the theory of evolution is still underway. Here, many still believe, traditional Christianity must make a stand. The churches opposed Darwin when he published the *Origin of Species* in 1859, and that opposition has shown no sign of relenting. In challenging Darwin, the British politician Benjamin Disraeli asked: "Is man an ape or an angel?" and famously replied that he was on the side of the angels (Monypenny and Buckle, 1929: 108). There are many who would still agree that we were created by God as described in the book of Genesis, not evolved from an ape by a process of natural selection.

Disraeli's quip is only one of the many skirmishes recorded in the battle over evolutionism. Even more famous is the clash between "Darwin's bulldog," Thomas Henry Huxley, and Bishop Samuel Wilberforce at the Oxford meeting of the British Association for the Advancement of Science in 1860. As the Darwinists remember it, Huxley demolished the bishop and cleared the way for Darwin to obtain a fair hearing. But his efforts came to naught sixty-five years later in the so-called "Monkey Trial" of John Thomas Scopes in Dayton, Tennessee. The result of that trial demonstrated that advocates of traditional religion were determined to protect the youth of America from the evolutionists' brand of materialism. Scoff as they might, liberals have been unable to hold back the tide of what soon became known as "creationist" opposition to Darwinism. And in some respects the opposition is quite justified, for modern atheistic Darwinists such as the biologist Richard Dawkins and the philosopher Daniel Dennett present the theory of natural selection as the final nail in the coffin of religious belief. They posit that if we are the products of blindly operating natural laws, any hope of seeing ourselves as the intended products of the Creator's will is out of the question.

In America, at least, the initiative seems to remain in the hands of the creationists. For several decades now, the Religious Right has maintained a constant opposition to the teaching of Darwinian evolutionism in the public schools. The young-earth version of creation science and more recently the idea of Intelligent Design (ID) are promoted as alternatives that must be taught to students. While I was writing this book, the Kansas State Board of Education debated whether or not alternatives to evolutionism should be included in the curriculum. In Kansas, the creationists are inspired by the Rev. Jerry Johnston of the First Family Church in Overland Park, who declared this an opportunity to reverse the country's moral decline. There was a similar, much publicized confrontation in 2005 in Dover, Pennsylvania, in which the creationists' claims were rejected by the courts. Scientists see the imposition of ID teaching as a recipe for undermining the quality of science education, thereby threatening jobs and economic security. More seriously, the journal *Science* published an editorial in April, 2005 warning that the latest attack on evolutionism might herald "twilight for the Enlightenment"—the final elimination of liberal thought from American life.

This image of confrontation between evolutionism and religion is so pervasive that to challenge it might seem quixotic. But the purpose of this book is to show that such a rigidly polarized model of the relationship benefits only those who want us to believe that no compromise is possible. I do not make this point because I subscribe to the compromise position myself—I am a pretty hardline skeptic on religious matters. But like Michael Ruse, I disagree with Dawkins and Dennett over the tactics to be adopted when confronted with the kind of situation that exists in America, or in any other country where fundamentalist religion tries to impose rigid limits on what scientists can investigate. Ruse is a philosopher of science who has played a major role himself in the controversies of the last several decades, defending evolutionism against the creationists' attacks. Yet in March 2005, he was reported as having disagreed openly with Dennett, who is perhaps the most aggressive Darwinist in modern America. Ruse argues that polarizing the situation further by stressing the most atheistic interpretation of Darwinism may put the whole enterprise of science and enlightenment at risk by inflaming the opposition. It may be better to oppose the fundamentalists by showing that they have oversimplified the response of religion to the quest for a science of origins. As a historian who has spent decades studying the response to Darwin, and as an observer of modern debates in America and Europe, I too believe that the best defense of evolutionism is to show the complexity of the religious approach to science. There are many scientists who still have deeply held religious beliefs, and many religious thinkers who are happy to accept evolution. Evolutionism is not necessarily atheistic, and creationism is not the only alternative open to the Christian.

To understand how this can be so, we shall make a survey of the history of the engagement of religious faith with scientific evolutionism, showing how a whole range of alternative positions have been explored, establishing a continuous spectrum of opinion where creationists and extreme Darwinists want us to see only black and white alternatives. Here a critical approach to history helps us to understand—if not resolve—the tensions that still divide the modern world.

THE USE AND MISUSE OF THE PAST

The debates sparked by Huxley and later by Scopes offer historical evidence that throws light on the cultural and social origins of the modern issues. But each of these episodes has become enmeshed in a web of interpretation that allows them to function as iconic images, exploited by those who have an interest in encouraging us to see the relationship between evolutionism and religion as polarized between two hostile camps. It is the Darwinists who have led the running in this effort to turn historical episodes into myths that help to shape our modern imagination. As the historian James R. Moore (1979) has shown, the metaphor of a war between science and religion was actually created by Huxley and his followers as part of their campaign to erect science as the new source of influence in modern society. Huxley's American disciple J. W. Draper encapsulated this interpretation of the relationship in his 1875 *History of the Conflict between Religion and Science*. The image of Huxley triumphing at the 1860 BAAS meeting is part of this mythology, designed to encourage the view that science reflects the freedom of the human intellect to challenge religious dogma. And the popular image of the Scopes trial reflected in the movie *Inherit the Wind* is part of the same ideology. Creationism is portrayed as a blind dogma incapable of standing up to the scrutiny of rational argument. These images are the stock in trade of the rationalists' argument against organized religion, in which the battle over evolutionism is merely an episode in a much wider campaign.

Modern historians have exposed the ways in which popular images of these events have been manipulated to create the myths that sustain the image of a war between science and religion. When eyewitness accounts written by those who were at the 1860 BAAS meeting are checked, there is little to support the view that Huxley forced Wilberforce to slink off in disgrace. Many of the biologists who endorsed the theory of evolution—including Huxley himself—did not accept Darwin's mechanism of natural selection as an explanation of how the process worked. Nor were late-nineteenth-century religious thinkers uniformly opposed to Darwin. Even in early-twentieth-century America, a careful survey of the tracts written by the fundamentalists shows that some of them were willing to accept a form of evolutionism. Several of the southern states refused to follow Tennessee in enacting legislation against teaching evolution in the schools.

These reinterpretations of key events in the story are part of a more general strategy in which historians have reassessed both the so-called Darwinian revolution and the overall relationship between science and religion. Darwin's strongly materialistic theory of natural selection did not begin to dominate biology until the early decades of the twentieth century. Rival theories presented evolution as a goal-directed system of progress, thereby evading the most dangerous implications of Darwinism. Perhaps the process of evolution has a divine purpose built into it. In recognizing this possibility we are led to a more general reassessment of the so-called "war" between science and religion. Through most of its history,

science has been undertaken by people who thought that by studying nature they were helping us to understand its Creator. It turns
out that this is true for many of the biologists who have developed
the modern theory of evolution.

Equally significant is the support for this less materialistic view
of evolution expressed by a wide variety of religious thinkers. James
Moore's book has helped to transform our view of the theological
debates over Darwinism by bringing to light the significant role
played by liberal religious thinkers hoping to bring the Christian
faith into line with modern attitudes and knowledge. It has to be
said, though, that historians have been less adventurous in seeking to uncover the complexity of the debate over evolutionism in
the early twentieth century. Here most historians' attention seems
to focus on the rise of fundamentalist opposition to Darwinism
and the events leading to the Scopes trial. Even those American historians who have transformed our understanding of the complex
events and attitudes surrounding the trial have written little on
the efforts of liberal Christians to create a synthesis with nonmaterialistic views of evolution.

Once we look beyond the evangelical religious movements that
led the assault on evolutionism in twentieth-century America, we
discover a very different world. American Christians of today may
be amazed to find out that some of their compatriots of a hundred
years ago welcomed evolutionism with open arms. Nor were the
liberals blind to the effect this would have on the basic tenets of
Christian faith. The idea of Original Sin was replaced by a faith in
the perfectibility of humankind under God's evolutionary plan.
In Europe, this liberal vision of Christianity did not even face the
rise of fundamentalist opposition to evolutionism that traumatized
America in the 1920s. Liberal religious thinkers were convinced
that they could make common cause with a science that had turned
its back on materialism. At a time when some American states were
passing laws forbidding the teaching of evolution altogether, the

Anglican clergyman and future bishop of Birmingham, Ernest William Barnes, hit the headlines by preaching what the London press called his "gorilla sermons" in Westminster Abbey. Barnes was trying to complete the synthesis of evolutionism and liberal theology begun half a century earlier. In so doing, however, he exposed cracks that had only been papered over in the earlier negotiations. If Christians accepted that humanity was the product of evolution—even assuming the process could be seen as the expression of the Creator's will—then the whole idea of Original Sin would have to be reinterpreted. Far from falling from an original state of grace in the Garden of Eden, we had risen gradually from our animal origins. And if there was no Sin from which we needed salvation, what was the purpose of Christ's agony on the cross? Christ became merely the perfect man who showed us what we could all hope to become when evolution finished its upward course. Small wonder that many conservative Christians—and not just the American fundamentalists—argued that such a transformation had destroyed the very foundations of their faith. Barnes had put his finger on a problem that still fuels the arguments of conservative Christians against evolutionism to this day.

Yet in America too the liberal view of Christianity was defended against the attacks of the fundamentalists. Famous preachers such as Harry Emerson Fosdick struggled to promote the flexible attitude to the biblical texts that had been developed in the previous century. This approach repudiated the idea of an inerrant text that had to be taken literally even on scientific matters and saw the Bible instead as a historical record of humanity's interaction with the divine. Significantly, though, Fosdick evaded the scientific debates on the cause of evolution and presented it as the unfolding of a divine plan toward its intended goal. Even this concession would not satisfy those who saw the Christian message as one of salvation for a sinful humanity, a vision that could never be reconciled with the ideology of progress. But scientists and liberal theologians continued to push the case for compromise, in effect following in the footsteps of Barnes and Fosdick. Historians are now beginning to explore the role played by liberal theology in the evolution debates of twentieth-century America.

At first sight one might have expected the modern opponents of Darwinism to welcome these historical initiatives. In fact, they show little interest in efforts to undermine the warfare metaphor, and seem indifferent to historians' efforts to create a more balanced view of the Darwinian revolution. They have a vested interest in maintaining the popular assumption that evolutionism can only be understood as a manifestation of atheistic materialism. Just like the atheists themselves, the evangelicals who endorse creationism want us to believe that hard-line Darwinism is the only form of evolutionary theory. If the materialistic Darwinians are the only true evolutionists, then evolutionism must be stopped if religious faith is to be preserved. But it is precisely this polarized image of evolutionism that has been undermined by historians' reinterpretation of the Darwinian revolution.

If Dawkins and Dennett can be seen as the intellectual heirs of Thomas Henry Huxley, the line that joins them has to be seen as only one strand in the complex web of interactions on the issue. But the atheists might argue that from the scientific perspective it is the most important position, because it is the only one compatible with modern biology. The combination of Darwinism and genetics has eliminated the non-Darwinian ideas of evolution that sustained the earlier hopes of a dialogue with religion. We can see this in the very different reactions of scientists and religious thinkers to one of the most charismatic mid-twentieth-century writers on this topic. Pierre Teilhard de Chardin was a paleontologist and a Catholic priest who achieved posthumous fame when his *Phenomenon of Man* was translated in 1959. The wave of enthusiasm for Teilhard's vision of humanity as the goal of the Creator's purposeful evolutionary process showed that the liberal tradition was still active. Yet by this time most scientists were suspicious. Teilhard offered only vague platitudes about how evolution worked, and this cut little ice with biologists who now saw the natural selection of genetic mutations as the only plausible mechanism. In this sense, the radical Darwinians are right to reject the liberal synthesis as a dead duck, because its scientific foundation is no longer plausible.

But the situation is not quite so simple. Ideas and attitudes are still developing, and liberal Christian thinkers are exploring ways of rendering the theories of Darwinian selection compatible with their faith. Bishop Barnes was in touch with the new Darwinism that was just beginning to emerge in the 1920s and 1930s—he knew Ronald Aylmer Fisher, one of the architects of the modern theory of natural selection who was himself a liberal Christian. Modern theologians who know their science—writers such as Arthur Peacocke and John Polkinghorne—seek a dialogue with biology in the full knowledge that it must include the Darwinian view of evolution. They explore ways in which natural selection can be seen as part of the Creator's purpose, even though it reveals that evolution has no central driving force aimed at a predetermined goal. If Dawkins and Dennett are T. H. Huxley's intellectual heirs, these thinkers are the heirs of Barnes and the earlier generations of liberal Christians who sought to accommodate the latest developments in science.

The liberal tradition in twentieth-century theology appears most visibly in European sources. This may be partly an artifact of historical analysis—as noted above, historians of American culture have tended to focus on the evangelicals' opposition to evolutionism. But this in itself reflects the interests of American culture in the late twentieth and early twenty-first centuries. Europe has only recently been exposed to the kind of evangelical opposition to evolutionism that has been characteristic of American religion since the 1920s. Europe is now a largely secular culture in which the most active form of religion is fundamentalism (both

Christian and Islamic) imported from abroad. The issues discussed in this book thus look very different when viewed from a European perspective. (I might add that for the last twenty-five years I have lived in Northern Ireland, one of the few areas in Europe where religion has retained a strong hold on the people, as a source of identity in a dangerously divided society.) But the liberal tradition is not absent from American religion, for all that it does not appear in the media, and a more balanced account of how religious thinkers have responded to Darwinism across the whole period since the *Origin of Species* was published may offer useful food for thought.

ISSUES THAT DIVIDE

Before launching into the reinterpretations sketched in above, it will be constructive to clarify the issues that define the debate. Far more is at stake than a simple confrontation between Darwinism and a literal reading of the book of Genesis. Evolutionism raises general issues about how God might govern the universe, and specific issues about the status of humanity within the universe and the wider scheme of creation. Within these two main categories there are a number of subissues, each of which can divide even religious thinkers who are conscientiously trying to articulate their faith in the face of the evidence offered by modern science.

It's also worth remembering that there are nonreligious traditions that share the creationists' distrust of Darwinism, but for very different reasons. Left-wing thinkers see the theory of natural selection as a means for articulating harsh policies of "social Darwinism" with an apparently scientific justification. But some of the values they identify with Darwinism are those shared by many Americans on the Religious Right. Such apparent paradoxes warn us that any attempt to understand the relationships between humanity and the natural world leads us into a minefield of rival value systems, all of which seek to justify themselves by discrediting their opponents' use of science.

There have been efforts to show that the whole debate is unnecessary and arises from a fundamental misunderstanding of the natures of science and religion. The noted paleontologist and science writer Stephen Jay Gould argued this in his *Rocks of Ages* (1999). His point was that science is concerned with facts, whereas religion deals with human values. There is no contact between the two enterprises because they are asking different kinds of questions. They are, in Gould's term, "non-overlapping magisteria," equally important but quite independent from one another. But to make this case Gould had to treat religions as nothing but ethical systems, and although it is true that all religions do endorse ethical values, they are much more than value systems. They seek to define the origin and nature of both humankind and the cosmos, and in most cases those definitions are derived from creation stories contained in sacred texts. To separate the creation myths from the value systems they support is to misunderstand the nature of religion, and here Gould's effort to cut the Gordian knot fails. Christians defend their values by defending a vision of how God created the universe, and that is why they cannot regard science as irrelevant. The question is: how rigidly does the belief system of a religion such as Christianity define the framework within which scientists can investigate the world?

To see why theologians and philosophers can fall out over how to deal with evolutionism, we must note that some of the issues are very general, in the sense that they would arise even for someone whose religious faith was not derived from a body of sacred literature. The philosophy known as deism postulates a God who designed the universe but took no further interest in it once He had created it. A deist has no interest in the creation story of the Bible (or of any other allegedly sacred text), but might still want to defend the idea that the universe shows some signs of being created by an intelligent Being. Such a philosophy is too impersonal for most religious believers: most traditional faiths are forms of theism, that is, they support the belief that God not only created the universe but also continues to take an interest in it. He may even interfere with its normal operations from time to time, such supernatural interventions being what we normally call miracles. One can be a theist in a general sense without accepting any of the existing theological traditions (or by combining elements from several of them, as in the Baha'i faith).

Turning to the traditional faiths, we shall be concerned almost exclusively with the various forms of Christianity, although the other great religions of the world have also taken positions on the issues raised by evolutionism. Some can be fairly relaxed about ideas that are deeply worrying for the great monotheistic faiths. Hinduism, for instance, has sacred texts which imply that the universe goes through great cycles of change over vast periods of time. It also refuses to make the clear distinction between humans and animals that seems so obvious to those religions that draw their origins in part from the Hebrew tradition.

There are three major monotheistic faiths that take what the Christians call the Old Testament seriously as divine revelation (this comprises the Jewish sacred texts, including the Torah, the first five books of the Old Testament supposed to have been written by Moses). Christianity adds to this the New Testament, which presents Jesus as the savior who will redeem us from the blight of Original Sin (Adam and Eve's disobedience that led to the expulsion of the human race from paradise). The Islamic faith accepts Jesus as a great prophet, but focuses its attention instead on the Koran, the revelation of the prophet Mohammed. Judaism and Christianity focus on the story of creation as described in the book of Genesis when they confront the alternative story of the earth's history told by modern geology and evolution theory. When taken literally, the story in Genesis implies that God made a single creation that included humans almost from the very beginning, and that creation has not changed since (except perhaps for the catastrophic events of Noah's flood). Humans are distinct from animals because only they were created with souls that will be judged by their Creator in some form of afterlife. One of the great problems evolutionism poses for this version of events is that it implies that we are derived by a gradual process from the animals, thereby casting doubts on the unique status of the soul.

The rest of this chapter provides only a skeleton outline of the relevant positions. Further details and guides to further reading are provided in the appropriate later chapters of this book. For general reading on the relationship between science and religion, see the classic texts by Ian G. Barbour (1966, 1968). Surveys of the history of the interaction between science and religion include Brooke, 1991; Ferngren, ed., 2002, and Lindberg and Numbers, eds., 1986, 2003. For more detailed surveys of the debates over evolution see Appleman, 2001; Durant, ed., 1985; Greene, 1959, and Moore, 1979, and for recent surveys of the issues raised by evolutionism, see Ruse, 2005.

THE SACRED TEXT

We begin with the problems posed by the appeal to a sacred text, in this case the Bible in general and the book of Genesis in particular. For many evangelical Christians, this is the great issue: if the Bible is the word of God, it must be taken seriously when judging any other account of the earth and humanity's origins. The Bible tells us that God formed the heavens and the earth in seven days, according to the first chapter of Genesis, with Adam and Eve being created on the sixth day (the seventh, of course, is the Sabbath). There is no mention of a significant period of prehistory (i.e., history before the appearance of humanity), and certainly no reference to periods in which the earth was populated by animals different to those we see around us today. The human race has existed since the creation, and the records allow us to date its origin, and hence by implication the creation of the universe itself. In the seventeenth century the archbishop of Armagh, James Ussher, added up the ages of the patriarchs mentioned in the Bible back as far as Adam and concluded that the earth was created in 4004 B.C.(at midday on Sunday, 23rd October, to be precise). The young-earth creationists of today have revived the view that the earth can be only a few thousands of years old.

Such an interpretation of the sacred record obviously rules out evolution, but it also rules out the whole package of modern sciences dealing with earth history, including geology, paleontology, and prehistoric archaeology. As critics of the young-earth position point out, it takes us back to a position that has not been taken seriously by working scientists since the late seventeenth century. The alternative creation science promoted by the young-earth movement revives the once popular idea that all the fossil-bearing rocks were laid down in Noah's flood, the one event mentioned in the Bible that might have completely reshaped the earth's surface. Significantly, the young-earth movement used to feel the need to offer an alternative *science* of the past, arguing that their theory can make better sense of the actual evidence from the rocks. Modern proponents of Intelligent Design also see their rejection of evolutionism as based on scientific arguments, although some creationists deny any authority to the scientific approach, claiming that the scientists are just rival storytellers trying to convince the audience by mere rhetoric.

Why do fundamentalists take the creation story literally? As we shall see, there are plenty of sincere Christians who are prepared to see creation in a metaphorical sense that is compatible with some form of evolutionism. Michael Ruse, himself an active participant in the debates on the side of evolutionism, explains in a recent study (2005) how the answer to this question lies in a particular vision of Christianity's predictions about the end of the world (which in the Book of Revelation will be preceded by the millennium, the thousand-year rule of Christ). Ruse argues that Christians can be divided into two camps, the postmillenarians, who believe that we can bring about the kingdom of God on this earth before the end, and the premillenarians, who think that nothing can improve this world and we should all be preparing for the coming of an external salvation. The postmillenarians are liberals who can be persuaded to take a more relaxed view on the word of Genesis. The premillenarians are fundamentalists who are forced to take the Bible story of the earth's origin seriously in order to defend their literal interpretation of the predictions about its end. The premillenarians are also opposed to the whole ideology of social progress, which they see as an illusion that distracts us from humanity's essentially sinful nature. Since evolutionism is often used to underpin the idea of progress, here is another reason for opposing a metaphorical reading of Genesis.

The young-earth version of the creation story takes the whole narrative literally, including the six days of creation, which are assumed to be days of twenty-four hours. But not all Christians take the word of God literally, at least in areas where it refers to matters of scientific fact. When defending his right to investigate Copernicus' theory that the earth goes around the sun, Galileo argued that the sacred record is not an astronomy textbook. Its purpose is to convey the Christian message of salvation to ordinary people, and it necessarily had to be expressed in language consistent with a common-sense worldview. It was written as though the earth were the center of the universe, because to raise the issues addressed by Copernicus would only confuse people over technicalities to no purpose, as far as the spiritual message was concerned. The sacred text was recorded by writers who—even though divinely inspired—could only relate events that were comprehensible to them and to their hearers at the time. To imagine that God's ability to create a universe was constrained by the level of scientific understanding achieved by the ancient Hebrews is to make a mockery of any notion of divine omnipotence. Note how Galileo's assumption takes for granted the idea of progress in human understanding of nature, paving the way for what Ruse calls the postmillenarianism of liberal Christianity. In the following century, the call to reinterpret the Bible on an increasingly wide range of issues generated the ideology of social progress that would challenge the structure of traditional Christianity—although this was the last thing Galileo intended.

The geological sciences soon provided evidence that the structure of the earth's crust is too complex to be explained as the product of Noah's flood. The evidence implied that there were extensive periods before humans appeared. As long as one accepted that the universe was divinely created in the beginning, then the actual wording of Genesis might not have to be taken literally on the details of how the earth was formed. Theological liberals argue that the Bible tells us about our origin as God's creatures, but it is not a geology textbook. There is no mention of dinosaurs and vast geological periods, because this would have confused the ordinary people who needed to be convinced of the moral heart of the story.

There are two ways of treating the text in an allegorical fashion. Perhaps the days of creation are metaphors for vast periods of geological time, each day representing a whole epoch such as the age of dinosaurs. It could be argued that there is some correspondence between the sequence of animal and plant creations mentioned in Genesis and that provided by the fossil record. Alternatively, the Genesis story seems to imply a gap between the initial act of creation of the universe and the more detailed story located in the Garden of Eden. Perhaps this gap included a vast period of time during which there were other creations not actually mentioned in the text.

If either of these interpretations is accepted, much of modern geology and paleontology can be accommodated. There would be a series of creations before the appearance of humankind. Perhaps Noah's flood was the last of many such catastrophes, each responsible for the extinction of whole populations. Some modern creationists accept this position and are even prepared to allow for a limited form of evolution in each period (including the early phase of the present world), as long as the ancestral form of each main type of living thing is presumed to be divinely created. The full evolutionary perspective rejects this compromise. Basing itself on a more general presumption that miracles do not occur in the world, it postulates that natural laws must be able to explain all of the developments revealed by the fossil record, up to and including the origin of humanity. Thus each new species has to be the modified descendant of an ancestral form, and humans must have evolved from an ape-like creature (since the apes are our closest biological relatives). This position does not necessarily rule out acceptance of miracles in the course of human history—it can be argued that although God does not normally interfere with His creation, He is willing to do so in order to focus our attention on the events that are crucial for our salvation. But the processes that shaped the development of the earth and its inhabitants should not be understood to include any supernatural interference. Without this presumption, the evolutionist argues, science cannot study these processes. Its methods cannot tackle the supernatural, and there would be no way of being sure where the realm of natural law ended and that of miracle began.

THE ARGUMENT FROM DESIGN

Once the decision has been made to adopt this evolutionary perspective, there are two main areas of concern for any religious thinker operating within the Judeo-Christian framework. The first is the question of design: if species are created by miracle, we know that they have been designed by a wise and benevolent Creator. Can we still believe that God has a hand in the creation of species if they have been formed by processes governed by natural law? This is not necessarily impossible if the laws themselves were instituted by God and govern a system that He intended to produce certain results. The second issue relates to the human soul. If—as the Judeo-Christian religions believe—humans are distinct from the other animals by virtue of possessing a spiritual element in their character, how is it possible for a species whose members possess such a unique character to have evolved gradually from one that does not? Evolution makes no room for a discontinuity: either the animals must have at least some primitive level of spirituality that could be enhanced, or the whole notion of the soul is a delusion.

Turning first to the question of design, the exponents of what is called "natural theology" suppose that in studying nature one is studying the handiwork of God and can expect to see evidence of His intelligence imprinted on what we see. The classic way of formulating the "argument from design" in the area of natural history is to demonstrate the complexity of the living body and the adaptations of its various functions to the necessities of life, and to insist that such a well-designed system cannot have originated by chance—it must be the direct product of the Creator's will. In William Paley's classic text *Natural Theology* of 1802 we find the analogy of the watch and the watchmaker. If we find a watch when walking through the countryside, argues Paley, we know that such a complex structure of springs, cogwheels, etc. cannot had been produced by the undirected forces of nature, and we presume that it is an artificial construct made by an intelligent person, the watchmaker. (I have never been sure whether or not this analogy still works with modern watches, which are just electronic "black boxes" as far as most of us are concerned.) By the same token, if we study the human or animal body and similarly find a complex series of structures all adapted to the end of keeping the body alive, we are entitled to suppose that undirected nature could not have formed it, and so here too there must be an intelligent designer, God.

Darwinists claim that natural selection can produce complex structures without the involvement of design, in effect by trial and error. New structures are built up by a process of tinkering, in which each slight improvement is preserved. In response to this challenge, modern creationists invoke Intelligent Design (ID) to preserve the essence of Paley's argument. The supporters of ID take the study of the living body onto new levels, investigating even the biochemical processes that keep the various functions operating. They claim to find evidence of complexity that rules out the possibility of intermediate stages by which evolution could have built up the structures. All parts of the system must function together, or it doesn't work at all. The only possible explanation involves some form of supernatural intelligence to design the whole system in a coordinated way. The Darwinists challenge the individual examples but also complain that to invoke the supernatural is to erect a barrier against any further scientific exploration of the topic.

History also poses a problem here, because there have been many religious biologists who were not impressed by this version of design. Paley focused on the usefulness of the structures possessed by particular species. Each animal has its own special features adapting it to its way of life. But to some biologists this seemed a rather crude notion of design, since it presented God as a kind of engineer, building a vast collection of individual structures according to no principle other than that of local expediency. They looked for patterns in the overall collection of living things, noting that underneath the individual adaptations there were relationships between species. The existence of these relationships hinted that the whole of creation formed a unified, harmonious design. Darwin's great opponent Richard Owen argued that all the vertebrates were superficial modifications of a single pattern, evidence that God designed them as a rationally ordered whole, not just a ragtag and bobtail of individual adaptations.

The problem with this argument is that once you start to see patterns linking species, it becomes much more plausible to imagine all the variations unfolding by a continuous process. Although widely dismissed as an opponent of Darwinism, Owen himself in the end came to adopt the idea of "theistic evolutionism." He believed that the emergence of individual species came about by the unfolding of a universal plan under the operation of natural laws that were expressions of the divine will. Here is the most obvious compromise between the idea of design and the theory of evolution. Evolution occurs, but it is not a totally natural process because the course of development, and the ultimate goal, is determined by God's designing intelligence operating within the laws of nature.

The problem with theistic evolutionism, as far as many scientists are concerned, is that we do not normally think of the laws of nature as entities capable of seeing and planning for the future. The law of gravity operates just the same whether you are sitting on an armchair or falling off a cliff—if it somehow modified itself to prevent a tragedy in the latter circumstance, it wouldn't be a real law. Theistic evolutionism was trying to incorporate the supernatural into the natural, leading the philosopher John Dewey to scoff at it as "design on the installment plan."

Biologists looked for a mechanism of evolution that would be lawlike in the manner normally accepted by scientists, but which would still allow them to believe that the universe was not a process of trial and error as Darwin had supposed. The most promising approach was known as "Lamarckism," after the French biologist Jean Baptiste Lamarck. Lamarckism works through a process in which animals improve themselves by their own efforts (see Chapter 2 for details). Lamarckism requires no struggle for existence and allows evolution to be led in a purposeful direction by the animals' recognition of what is good for them. This sounds like the kind of process that a wise and benevolent God would institute as a means of allowing His creatures to flourish in the world. Yet it works by what appears to be a combination of perfectly natural processes.

The only problem with Lamarckism was that by the early twentieth century, the science of heredity had shown that characters acquired through an animal's efforts cannot be inherited. The genes pass on characters in a predetermined manner and cannot be affected by changes in the organism that carries them. The only alternative mechanism of adaptive change is Darwinian natural selection, which can fairly easily be adapted to the genetic model of heredity. And here we see the central importance of Darwin's theory in this debate, because natural selection does not look at all like the kind of mechanism a wise and benevolent God would institute to bring about adaptive evolution.

There are two reasons for this. The first is that the raw material of selection is the minute variations that allow each organism (like each human being) to be recognized as an individual. These variations are sometimes said to be random, not because they are uncaused, but because whatever causes them seems to have no regard to what would be beneficial to the individual or to the species. People have all sorts of different hair colors, and it doesn't seem to make any difference to their lives. In modern genetics, this variation is seen as the result of genetic differences. Ultimately this range of genetic variation is caused by mutation, a form of copying error in which a gene that used to code for a particular character is changed so that it produces something different. And precisely because these are copying *errors*, they do not appear according to the needs of the individual, and many of them are positively harmful. The raw material of natural selection has no built-in design, no way of anticipating the future needs of the species: it is a process of trial and error. The God who chose to create a universe in which evolution worked in this way was certainly not taking a hands-on approach.

The reason why a chaos of original variation can produce characters that look as though they have been designed is the process of selection. In any new environment, genes that code for what has now become a useful character will increase their frequency in the population, because the organisms which carry them will breed more readily. Those with maladaptive genes will not do very well in what Darwin called the "struggle for existence" and will not breed—they may well die. The proportion of genes conferring adaptive benefits thus increases and the species evolves toward an appropriate specialized character.

Here is the second reason why theologians have found it hard to accept Darwinism as a mechanism instituted by God—the whole process is driven by death and suffering. To be fair, this isn't a problem for the theory of natural selection alone. Darwin and his followers have provided enough evidence to show that there must be a massive elimination of individuals within every population just to keep the numbers stable. The basis for what Darwin called the "struggle for existence" is built into nature—whether or not it serves as the driving force of evolution. Many Christians find the notion of a world governed by struggle and suffering as abhorrent, although some biblical literalists see it as a consequence of Original Sin. On this model there was no struggle before the Fall, and the literalists are disturbed by paleontologists' claim that the fossil record shows the prevalence of death and predation in the animal kingdom long before humans appeared.

Liberal theologians accept a role for struggle and suffering, seeing it as a creative agent, a process that encourages us all to better ourselves. This is hardly a valid model for the Darwinian theory, but in the latest versions of liberal theology a more realistic effort is being made to accommodate the harsher side of nature. Theologians point out that Christianity is unique among religions in seeing suffering as an integral part of the relationship between the human and the divine. Suffering and conflict are inevitable in a world blighted by sin, and Christ's suffering on the cross—the price of salvation—allows the divinity to participate in this aspect of the world's operations. Perhaps, then, it should not surprise us to find that suffering is in fact part of the creative process by which we were formed. Paradoxically, Christianity may be in a better position to deal with an evolution theory based on the struggle for existence than other religions, which take a less pessimistic view of the human situation.

HUMAN ORIGINS

The question of human sinfulness points us toward the other major area of concern for religious thinkers confronting the challenge of evolutionism. Whatever the mechanism of change, evolution presupposes that humans have evolved from animals, with the chimpanzee as our closest relative. This should not be a problem for a religion such as Hinduism, which accepts the possibility that souls now inhabiting human bodies may be reincarnated in animals in some future life. But Christianity belongs to a group of religions which base their beliefs on texts stating that humans were created with intellectual, moral, and spiritual powers transcending the mentality of animals. For Christians, then, the idea that humans with immortal souls have emerged by a gradual process from the "brutes that perish" is deeply disturbing. How can a natural process have produced these higher levels of existence from so unpromising a raw material? Isn't it obvious that the soul must have been specially created, appearing only in the first humans? And in this case, doesn't it make more sense to believe that the first humans—Adam and Eve, if Genesis is taken literally—were created body and soul by a miraculous act of God?

The problem is compounded by Darwinism's focus on the survival of the fittest as the mechanism of change. If species only change when populations adapt to new conditions, there is no necessity for evolution to be progressive and no possibility of seeing humans as the goal of a predetermined plan. The old idea of a ladder of creation with humans at the top allowed the religious believer to interpret evolution as the unfolding of a divine plan that had humanity as its ultimate goal. Darwinism turns the ladder into an ever-branching tree in which no one branch can be privileged as the main trunk, no final twig as the goal of creation. And if natural selection is the process of change, the motor of evolution has no purpose—it is a totally amoral sorting of the best-adapted individuals generated by random genetic mutation. How could such a mechanism produce the moral and spiritual characters that some Christians believe raise us above the animals?

Darwin himself tried to make the case for the *Descent of Man* (the title of his 1871 book on the topic) by minimizing the gap between the higher animals and the "lowest" humans. By modern evaluations, he exaggerated the mental powers of animals by accepting anecdotal evidence of their intelligence and even their moral awareness. He also depicted some living races of humanity as closer to the ancestral ape, in a manner we would find quite unacceptable today. Modern scientists are well aware that however small the genetic difference between humans and chimpanzees, humans do indeed have mental faculties that are significantly more advanced than those of even our closest relatives.

Perhaps the greatest point of controversy centers on our moral sense or conscience. Darwin tried to explain this in terms of our social instincts, implanted by evolution in any species in which the individuals live in cooperating groups. The more militant of the modern Darwinians are only too happy to rise to this challenge by insisting that the Christian view of the human situation is fundamentally unrealistic. We are, they insist, only improved animals, still driven by animal desires despite our increased intelligence. Modern evolutionary psychology is seeking ways of explaining how the various faculties of the mind have evolved in the circumstances to which our immediately prehuman ancestors were exposed. The science of sociobiology explains all animal behavior, including human social behavior, in terms of instincts generated by natural selection acting among groups of genetically related individuals. Morality is just another product of what Dawkins calls the "selfish gene."

Modern creationists often accuse Darwinism of encouraging us to behave brutally to one another. After all, the theory does tell us that we are no better than brutes, so we should not be surprised if it is used to argue that our behavior is programmed to include brutal instincts. They talk darkly of the horrors of social Darwinism, and point to Nazi Germany to illustrate what happens when political leaders glorify the struggle for existence. But the creationists are usually silent on the ideological origins of Darwin's theory, which historians link to the free-enterprise culture of Victorian Britain. The political Left dislikes social Darwinism too—but its preferred example of unrestrained struggle is the competitive individualism of the capitalist system. By this standard, it is the free-enterprise ideology favored by most American creationists that counts as social Darwinism!

This issue warns us of the need to be very careful in assessing the alleged implications of the claim that humans are governed by biological instincts. There have been many different forms of social Darwinism, depending on whether the struggle for existence was seen as operating between individuals or groups (e.g., nations or races). Hitler certainly pointed to Darwinism as one source of his vision of nations locked into a struggle for supremacy, but many opponents of free-enterprise capitalism have seen that political system as the more natural analog of Darwinian biology. To many Americans, the free-enterprise system seems the guarantee of freedom and economic progress, but they would do well to recognize that when Darwinism first appeared, it was precisely the hope of this form of progress that encouraged many to support it.

To creationists the analogy between free-enterprise capitalism and the Darwinian struggle for existence seems absurd. The whole point of their rejection of evolutionism is to defend the claim that the human spirit is something lifted above the level of brute nature. Self-reliance and the drive for personal success should always be tempered by Christian values, which can have no basis in animal behavior. This brings us back to the central problem posed by Darwinism, its implication that human nature is simply an improved version of the mentality of animals. How can the higher moral values emerge from a brutal struggle for existence? The simple answer for creationists is that they cannot, and hence we need to see humans as the products of supernatural creation, not of natural processes. But the analogy noted above between Darwinian struggle and free-enterprise individualism reminds us that it is not always easy to define what is part of nature and what rises above it. Conservative religious thinkers have always tended to take the hard-line position against evolution. But liberal thinkers have tried to find a way of accepting that we may be the product of nature, while portraying nature as something capable of lifting its products steadily up toward higher things.

History shows us that there are many ways of trying to soften the impact of evolutionism on the Christian view of human origins, although they usually involve modifying the central tenets of Darwinism. One obvious tactic, still the official position of the Roman Catholic Church, is to accept the evolution of the human body from some lower form but insist that the soul was an entirely new entity created and miraculously implanted in the first true humans. Most Darwinists see this as pointless: why go to all the trouble of formulating a comprehensive theory of evolution only to concede that it does not apply to the most interesting and original development in the history of life on earth? A more promising approach is the idea of emergent evolution. This assumes that evolution is continuous at one level, but occasionally reaches thresholds or breakthrough points where something entirely new enters into the world. The appearance of the human mind would be one such breakthrough. The Darwinist, however, still looks for the causal mechanisms that create the new faculties, and tends to find the notion of emergence a meaningless concession to outdated religious preconceptions.

Another approach is to focus on the whole pattern of evolution, in the hope of seeing evidence that the human mind is the intended product of a process instituted by the Creator. In the nineteenth century it was widely assumed that evolution was inevitably progressive. The tree of life was routinely depicted with a central trunk that ran up toward humankind as the goal of creation. This was why the social Darwinists assumed that the struggle for existence was the motor of both biological and social progress. Liberal religious thinkers also took comfort in the idea of progress, seeing it as evidence that the whole evolutionary process represented a divine plan driven by mechanisms that were inherently purposeful. To them, it did not seem quite so unreasonable to imagine that even the higher human faculties were produced by such a process. The modern Darwinian perspective (seldom fully appreciated by Darwin's immediate followers) makes this assumption more difficult to sustain, because there is no goal toward which evolution is moving, and the mechanism of change is anything but purposeful. One of the greatest challenges for those present-day theologians who wish to engage with Darwinism is that presented by evolutionary psychology's efforts to explain human behavior as driven by mechanistic processes in the brain, established by natural selection.

For the theologians to deal with this issue, they have to confront the problems identified in Barnes's gorilla sermons. The problem with linking evolution to the idea of progress is that by turning humans into the goal of evolution, we imply that human history is only a continuation of the advance that has taken place through the animal kingdom. But Christians have traditionally assumed that history is not progressive: humans have fallen from an original state of grace through Original Sin, and can only gain salvation through Christ's sacrifice on the cross. To argue—as Barnes and his modern successors must do—that we have risen from the apes as part of God's plan is to miss the point of Christianity's belief that we are contaminated by sin, that the divine purpose has been frustrated by humanity's willful separation from God after its creation. This is the basis of the premillenarianism of the fundamentalist position identified in Ruse's analysis: Evolution is false not only because it denies that we are created by God, but because it is linked to an ideology of social progress, which claims that we can improve conditions here on earth. Liberal Christians may adopt the postmillenarian position, in which we bring about the kingdom of God through our own efforts before the end of the world, but for the evangelical the only hope for sinful humanity is salvation through the acceptance of Christ.

Here again, though, the Darwinian emphasis on the undirected nature of evolution may turn out to be a hitherto unrecognized advantage. Perhaps the old-fashioned liberals were wrong to emphasize progress in order to see us as products of a rigidly preordained divine plan. If we see evolution as a more experimental process, groping its way upward against all the odds, we can better understand the tensions that lie at the heart of the human situation. We are animals who have acquired higher powers by what the atheist sees as a cosmic accident, but which the Christian might understand as the Creator's only way of producing beings with the freedom and the ability to challenge their biological inheritance. The very fact that Christianity takes such a pessimistic view of the human situation makes it the best-placed of all the major religions to deal with the challenges of Darwinism. For those Christians who can face the prospect of breaking with a literal interpretation of Genesis, the fact that evolution does not seem to be focused on progress and preordained purpose offers a chance to explore the possibility of a creative synthesis with modern biology. In recognizing that nature is not so obviously designed as the natural theologians imagined, we see that the Christian sense of the imperfection of humanity was not misplaced after all.

A historical study of the encounter between evolutionism and religion may thus pave the way for a better understanding of the tensions so obvious in the modern world. The story has not been one of endless conflict. It has also involved efforts to establish a synthesis that have required both sides to think carefully about underlying principles. The literalist will never compromise, of course, and we need to understand why. But the liberal position has itself evolved over time, and is still evolving today.

CHAPTER TWO

SETTING THE SCENE

In the conventional account of the Darwinian revolution, the *Origin of Species* assaults an unsuspecting public still convinced that the world had been created as described in Genesis. Up to 1859, so the story goes, everyone assumed that all the animals and plants (including the human species) had been created by a wise and benevolent God sometime around 4004 B.C. Darwin's radical theory was a bombshell lobbed into a world dominated by traditional Christian beliefs. Supported by a strident band of materialists led by firebrands such as Thomas Henry Huxley (known as "Darwin's bulldog"), the theory of natural selection was used to establish a worldview dominated by chance and conflict, rather than design and harmony. Soon the stability of the social order was being disrupted by an ideology of social Darwinism in which the strong justified their exploitation of the weak by claiming that life and death struggle was "only natural." Western society has been living with the consequences of this cultural and ideological revolution ever since.

The history of science shows that this model of sudden transformation is very far from the truth. By the time Darwin published, many aspects of the old worldview had already been abandoned, at least by those with any knowledge of science. The geologists had established that the earth was very old (although not as old as scientists now think today) and had uncovered a fossil record that included a sequence of animal populations succeeding each other through this vast period of time. Those who wanted to preserve natural theology had to update it in the light of these new discoveries. If it was no longer possible to imagine God creating just the modern species in the Garden of Eden, perhaps He had made a series of creations at the start of each geological period. Many liberal Christians continued to endorse a more sophisticated natural theology as a way of affirming that the Creator cared about the universe.

We also need to be aware of the complexity of Christian responses to science. Evangelicals saw inherent limitations in natural theology—it might confirm the existence of God but it could not reveal the route to salvation through Christ. Only revelation could do that. But this did not mean that the study of nature was wasted, since the universe was still a divine creation and it must be possible to reconcile the truths revealed by the book of nature and the book of divine revelation, the Bible. Evangelicals were worried that the new developments in science might lead the unwary to question Christianity, and they worked hard to provide an acceptable interpretation for ordinary readers (Fyfe, 2004).

What the religious writers on science were worried about was the opportunity the new discoveries provided for those campaigning against formal Christianity. Radical thinkers began to explore the possibility that some form of natural process had produced the succession of forms revealed by the fossil record. The crucial questions were whether or not this idea could be reconciled with the belief that the universe was a divine contrivance, and whether or not humans could be seen as the product of a divinely planned sequence of development. Some of the ideas thrown up at this time have a superficial resemblance to the modern theory of evolution, although historians such as Jon Hodge (2005) argue that it is misleading to use the term "evolution" when discussing these early ideas. To use the modern term is to invite misinterpretation of theories that were based on foundations virtually incomprehensible to a twenty-first-century biologist. There is something to be said for this argument, but for the sake of simplicity, I shall not be following Hodge's advice in this chapter—although I shall where possible use alternative terms such as "transformism," which came into use before the modern meaning of "evolution" was formulated.

By the time Darwin published the *Origin*, the basic idea of transformism was being widely discussed, although not necessarily in the radical form represented by the theory of natural selection. Some liberal Christians accepted that new species were modified from old ones, but did not see this as a foundation for a materialist worldview. They believed that if new species were created by natural law rather than divine miracle, the laws themselves were nevertheless established by the Creator. The outcome of the laws' activity could still be seen as the expression of God's will, with the human race as the goal toward which the whole process was aimed. This was the position advocated in Robert Chambers's popular bestseller *Vestiges of the Natural History of Creation*, published in 1844 (fifteen years before the *Origin of Species*).

Although condemned by conservative thinkers, *Vestiges* got everyone talking about the idea of evolution, and according to historian James Secord (2000), had persuaded many to take the idea seriously. Secord suggests that we should see Darwin as merely completing the revolution that Chambers had initiated. This is an interesting point, especially given the fact that Darwinism was often linked with the idea of progress. *Vestiges* shaped the way in which Darwin's book was read, leading many to see it in a light very different to that in which we perceive it today. My own feeling is that we still need to see the *Origin of Species* as a deeply influential text. This is partly because it does contain within it the ideas on which many of the most radical aspects of modern evolutionism are based. But even in the 1860s, Darwin's theory played a vital role, convincing many working scientists that evolution was a respectable theory that they could endorse, not just a vague metaphysical speculation.

This chapter will fill in the details of the scientific and cultural transformations that had taken place in the pre-*Origin* period, helping us to get a sense of just how far things had gone before the theory of natural selection first raised its head. We shall see how the geologists established the science of the earth's origins, creating the system of geological periods we still accept today. We shall also see how radical thinkers had begun to challenge the vision of natural theology by arguing for theories of transformism with some resemblance to modern evolutionism. But we shall also have to confront the fact that some liberal Christians had already begun to accept a theology in which God formed the world by law rather than miracle.

NATURAL THEOLOGY

The assumption that the world has been created by an all-wise Creator was dominant in the seventeenth century. It remained popular into Darwin's own time, although by then it had begun to face challenges both from the radical thinkers of the eighteenth-century "Age of Enlightenment" and by naturalists trying to make sense of the evidence revealed by the fossils and the complex rock formations of the earth's surface. Despite these challenges, William Paley's *Natural Theology* of 1802 was still presenting arguments very similar to those of John Ray's *Wisdom of God in the Creation* published over a century earlier. The assumption that the structure of each species confirms the wisdom and benevolence of its Creator offered a powerful incentive for religious naturalists anxious to justify the application of scientific techniques to the structure of the earth and its inhabitants. By 1802, however, the assumption that the world had been created fairly recently was already beginning to lose its credibility.

The worldview still promoted by modern "young-earth" creationists was formulated by the naturalists of the seventeenth century. It was James Ussher, Archbishop of Armagh, who calculated that the earth had been created in 4004 B.C. (on Sunday 23rd October to be precise), an estimate that was soon being included in the margins of many Protestant Bibles. Ussher was no scientist, and his chronology was part of an active theological debate in which he was a major participant. Unlike the early Church Fathers, on whom the Roman Catholic Church relied for its interpretations of the Bible, the Protestant theologians were obliged by the principles of the Reformation to approach the text of the Bible as a divine revelation that every Christian was responsible for reading for him or herself. This encouraged them to take literally the accounts of creation in the Book of Genesis, so it did not seem unreasonable to assume that the creation of the universe immediately preceded the creation of the first humans. To establish when Adam was created Ussher had to engage in a complex scholarly process in which the Bible was correlated with other ancient records. He is often ridiculed for merely counting back through the generations of the patriarchs recorded in the Bible until he got back to Adam, but in fact there were many other factors that had to be taken into account. The exact dates he specified were dictated by his position in controversies of the time and involved complex symbolic interpretations of the text. But once the creation of Adam was dated, the lack of any sense of prehistory meant that it was then only a matter of adding on six days to get back to the creation of the whole universe.

Ussher's basic worldview was taken almost for granted by most of his contemporaries. Whether or not they accepted his exact dates, they presumed that the world had been created directly by God, and they had no reason to suppose that this occurred outside the timescale of human history recorded in the Bible. Thus early naturalists studying the rocks and fossils almost inevitably began from a young-earth position. As we shall see in the next section, this soon began to crumble. But the idea that all the species of animals and plants (including humans) had been created by God proved more robust, serving as a foundation for much natural history through into Darwin's own time. The assumption that species were divinely created encouraged the belief that all forms of life should show clear evidence of their supernatural origin. If they were created by a wise and benevolent God, we should expect them to be carefully designed both in their internal structure and in the ways that structure adapted them to the environment in which they live.

One of the most influential naturalists to develop this argument was the English naturalist John Ray. Although the son of a blacksmith, Ray was able to study at the University of Cambridge (Raven, 1942; Greene, 1959, chap. 1). He became an important figure in the scientific community, which included such luminaries as Isaac Newton and Robert Boyle. He helped to engineer a major transition in how naturalists approached the study of animals and plants. In the Middle Ages and the Renaissance, each species had invariably been described along with its symbolic role in heraldry, mythology, and the Bible. But by the end of the seventeenth century, the new scientific method proclaimed the importance of purely factual information. Naturalists like Ray began to insist that each species should be described and classified solely in terms of its physical characteristics and its relationship to the environment. The historian Peter Harrison (1998) has argued that this apparently modern position was inspired by the Reformation, which had established the Protestant churches in opposition to Roman Catholicism. The Protestants, by insisting that everyone should read the Bible for themselves, stripped the sacred text of all the symbolic interpretations added to it by Catholic theologians over the ages. This, argues Harrison, encouraged naturalists to take a similar position on the study of God's creation. It too was stripped of its traditional symbolism, becoming just a physical system designed by God, to be described solely in terms of the structures He had created.

We know that Ray was a deeply religious man who saw the world he described as a divine creation. He is often described as a Puritan who eventually gave up his fellowship at Cambridge rather than subscribe to the Act of Uniformity imposed after the restoration of the monarchy under Charles II (the Puritans had sided with Parliament in the English civil war). Recent work by Susan McMahon (1999) shows, however, that he was more inclined to the liberal Anglicanism favored by the Restoration government. He may have given up his fellowship to provide more opportunity for studying natural history under the patronage of his wealthy friend Francis Willughby. Ray's science thus fitted into the ideology promoted by Newton, Boyle, and the other founders of the Royal Society of London, who used their vision of a divinely created universe to bolster the new political order that replaced the chaos of the civil war and its aftermath.

Following this new policy, Ray set out to describe and classify the species available to him. The process of classification was itself seen as a process of uncovering God's plan of creation. We can see varying degrees of similarity between the species we observe. Thus the lion, tiger, and leopard are all "big cats" and can be grouped together in what would later be called a genus. Fifty years after Ray's pioneering efforts, the Swedish naturalist Carl Linnaeus introduced the modern technique of using the nomenclature (the naming system) to help describe these relationships (J. Larson, 1971). Each species was given two names, the first indicating its genus, the second the individual species. Thus the lion was *Panthera leo,* the tiger *Panthera tigris,* and so on. There are also deeper relationships—the big cats are all carnivores (along with the dog family, etc.) and the carnivores are mammals (along with all the other warm-blooded animals that suckle their young). Today, these degrees of similarity are seen as a product of common descent. The lion, tiger, and leopard are all very similar because they have evolved from a common "big cat" ancestor and still retain its basic character. But for Ray and

Linnaeus, the relationships were an indication that the world was an orderly system designed by a rational Creator. Classifying species was itself a revelation of design in the world.

But Ray's most influential approach to design turned in another direction and focused instead on the actual structure of individual species. His expression of what is called the "argument from design" anticipates the position of the modern advocates of Intelligent Design. He saw each species as a well-designed piece of engineering, superbly crafted to function in a particular environment. For Ray, there was no way in which unaided nature could have produced such intricate mechanisms. Design implied a designer, and the only possible designer was God. Revealing the purposeful complexity of natural structures provided an argument for the existence of an intelligent Creator. This was the theme expounded in his most influential book, his *Wisdom of God in the Creation* of 1691. Here he saw the earth and the heavens as physical structures designed by God to allow a habitat fit for living things in general and human beings in particular. As a naturalist he focused most attention on the various species of living things. The human body provides the clearest examples of design—most obviously the eye and the hand. The eye is a complex optical instrument, more flexible than the microscope and the telescope, which were revolutionizing the science of the time. The hand shows a complex system of bones and muscles without which our use of tools would be impossible. For Ray, these and many other examples showed that the human body had been designed by a Creator of infinite wisdom, but also of great benevolence, because every one of these complex mechanisms was an adaptation to our needs. How could such a complex set of mechanisms have been produced by a natural process—especially if the world were only a few thousand years old?

As a working naturalist, however, Ray did not confine himself to the human body. He included a whole catalogue of animal species, each of which could be shown to have its own equally well-designed set of organs and adaptations. The tiger has the teeth and claws it needs to function as a predator, and is camouflaged so it can stalk its prey effectively. And in case anyone thought that predators were incompatible with divine benevolence, the advocates of natural theology pointed out that they only killed the aged and sick members of the prey species, thereby saving them a long and lingering death. This whole approach represents what may be called a "utilitarian" argument from design. It shows that nature is designed by a creative Intelligence, and sees the best illustrations of that wisdom in the usefulness of all the individual structures. Ray and his followers introduced a focus on the adaptation of structure to function that would serve as a model for Darwin over a century later— although Darwin looked for a natural as opposed to a supernatural explanation of how such structures could be formed.

Ray's argument came under fire from skeptics during the eighteenth century, but it was still intact as far as religious thinkers were concerned over a century later. It was expounded with equal force and ingenuity by the Anglican clergyman William Paley in his *Natural Theology* of 1802 (Le Mahieu, 1976). Paley was an influential figure who became archdeacon of Carlisle and wrote extensively on both religious and social issues. He linked the two by arguing that the free-enterprise system gave a natural outlet for the faculties with which God had endowed the human mind. Some of Paley's books were used as textbooks at the University of Cambridge, and it was here that the young Darwin read his *Natural Theology* and was entranced by its account of all the adaptive features displayed by the various species of animals. It was Paley who popularized the best-known analogy used to support the argument from design (although it had actually been in use since the seventeenth century). If you were walking along and found a watch by the wayside, you would immediately know that you were dealing with something made by a human being. Looking at the watch, you would see a complex network of gears, springs, etc. all adapted to the purpose of driving the hands around the face to tell the time (remember that Paley was dealing with old-fashioned clockwork, where you can actually see the workings in operation). Nothing in nature could have produced this elaborate complexity: the watch implies the existence of a watchmaker, an intelligent craftsman who designed and built it. Surely the same is true, Paley argued, for the complex structures of living things. They are carefully engineered to provide the animals with structures that will allow them to function in the environment in which they live. Their structure also implies a designer, and the only conceivable Designer in this case is a wise and benevolent God. It is Paley's analogy with the watch and the watchmaker that provides the inspiration for the title of Richard Dawkins's book *The Blind Watchmaker* (1986), because Dawkins thinks that Darwin provided us with a theory (natural selection) in which blind nature—nature unaided by any supernatural designer—can produce these adaptive structures.

Paley's argument reduces God to the status of a brilliant engineer, bearing little resemblance to the Christian God of judgment and redemption. Those who preached a more emotionally inspired version of Christianity were inclined to be suspicious of natural theology. For the evangelicals who became increasingly active in the late eighteenth and early nineteenth centuries, the established churches had become only too willing to evade the true message of the Bible, which is that humanity is inherently sinful and can only be redeemed through Christ's sacrifice. By allowing the scientific study of nature to take precedence over the Bible's message, the natural theologians were opening the door to a materialism that would ultimately seek to replace the Christian message altogether. As we shall see, the exponents of natural theology had to work hard to ensure that their position did not become corrupted in this way. Some more extreme evangelical groups certainly turned their backs on the new science altogether. Recent historical studies have shown that in this earlier period there were many evangelicals who still hoped that better study of the book of nature would show that it could be reconciled with the Bible, even if it told us nothing about the message of salvation (Fyfe, 2004 and more generally on evangelicals and science, Livingstone, Hart, and Noll, 1999). Their popular literature provided an antidote to the materialist ideas being promulgated by radical groups. But much of the popular evangelical literature did not engage very closely with the latest developments, leaving its readers with a comfortable but highly vulnerable sense of security.

THE AGE OF THE EARTH

In 1691 Ray could still write as though there had been just a single creation. But even he was already becoming concerned by evidence from the study of rocks and fossils. Much of the earth's surface is composed of rock formations that look as though they have been laid down by natural forces in the course of what was probably a long period of time. There are stratified rocks that appear to be layers of hardened mud or sand, and these are often folded and twisted, and sometimes interspersed with veins of volcanic larva. There must have been great changes to the surface after these rocks were laid down. More seriously still, the sedimentary rocks contained what appeared to be the petrified remains of once-living things, some of them quite unlike the animals and plants of the modern world. Many naturalists at first attributed this evidence of change to the effects of Noah's flood. But this did not really explain why the fossils seemed to indicate a sequence of different populations succeeding one another in the course of time. By the time Paley wrote his *Natural Theology* in 1802, it was becoming hard to resist the implication that there had been not one creation, but many, spread over a period of time far greater than that allowed for by Archbishop Ussher's calculations (Greene, 1959; Haber, 1959; Rudwick, 1972, 2005).

In the late seventeenth century, some of Ray's contemporaries tried to salvage the traditional timescale by arguing that the rocks were all the product of the global catastrophe recorded in the Bible, Noah's flood. Thomas Burnet's *Sacred Theory of the Earth* of 1681 supposed that the mountains were formed when the rocks of the original crust collapsed into underground caverns. The water displaced from the depths generated the great flood and then settled down to become the oceans of today. William Whiston, a follower of Newton, suggested that the flood was caused by a comet that nearly collided with the earth. Burnet and Whiston were sufficiently in tune with the new scientific worldview to suggest that the flood was a natural, not a supernatural event. Burnet argued that an all-knowing God could foresee the sins of Noah's generation and set up the machinery of nature so that it would result in a flood at the appropriate time to serve as punishment. Even so, he was criticized by clergymen for playing fast and loose with the text of Genesis.

Invoking a flood that covered the whole earth could explain how fossil-bearing rocks could be formed even high in the mountains. But as Nicholas Steno and Robert Hooke had noted as early as the 1660s, there was evidence of massive earth movement *after* the sedimentary rocks were laid down. Steno also realized that there was a sequence in which the sedimentary rocks were formed, with the lowest being necessarily older than those which now lie on top of them. Neither dared to challenge the traditional biblical timescale, and were thus forced to imagine great catastrophes in addition to the flood that reworked the surface very rapidly.

Hooke and Steno were convinced that fossils were the remains of once-living things buried in sediment which had now hardened into rock. Steno was an anatomist, born in Scandinavia, who converted to Catholicism when he moved to work in Italy. He studied a shark cast up on the Italian coast and showed that its teeth were identical to a common type of fossil. Fossils were not a problem for those who saw the flood as the source of the sediment that formed the rocks. But Steno knew that the lower rocks contained different fossils to those found in the overlying rocks. Perhaps this was due to differences in the rate at which the corpses of the various forms of life settled to the bottom of the flood water (the explanation still used by young-earth creationists today). There was an alternative possibility, however, which seemed more plausible once it was recognized that there were major earth movements after some of the sediments were laid down. Perhaps the deeper rocks were significantly older than the superficial ones. There may have been successive periods of rock formation, an idea that would become increasingly difficult to reconcile with a short timescale. Hooke pointed out that some of the older fossils—the coiled shells known as ammonites, for instance—were different to any species known in the world today. He thus confronted a major problem for the idea of divine creation: the possibility that some species had gone extinct in the course of the earth's history.

In the course of the next century or so, the majority of working naturalists became convinced that the structure of the earth's surface could not be explained by a single catastrophe even such as the flood. There was just too much evidence of major changes—earth movements and erosion—occurring after the older rocks were laid down. Something like the modern idea of a sequence of geological periods began to take shape, and the timescale began to expand beyond that allowed by Ussher. In 1749 the radical French naturalist, the comte de Buffon, suggested that the earth might be 70,000 years old—trivial by modern standards, but a tenfold increase on Ussher's figure (Roger, 1997). Buffon was the superintendent of the French king's botanical garden in Paris, an influential position that gave him considerable immunity from ecclesiastical interference. His estimate of the earth's age was indeed criticized by the Church, but he issued only a token response and returned to the topic in 1778 to postulate a sequence of what he called the "epochs of nature," each characterized by a different set of conditions. Buffon thought the earth had been knocked off the sun by a colliding comet, so he was convinced that the surface temperature in each period had been successively lower. He also thought the oceans were gradually getting shallower, which explained fossil-bearing rocks in mountainous regions without recourse to either a flood or a catastrophic earthquake. In his theory there were seven epochs, in the course of which the earth gradually took on its modern form and the environment became suitable for the species we know today. In effect, Buffon's epochs were what we now call geological periods. The somewhat arbitrary division of a fairly gradual process into seven epochs was driven by the suggestion that those wishing to retain some connection with the biblical record could see each epoch as one of the "days" of creation.

Buffon was creating a true history for the earth, although its stages were defined by a cosmological theory in which the planet steadily cooled down. He also saw that one could appeal to evidence derived from the rocks to establish the sequence of events. The development of what eventually became known as the science of geology came from ever more detailed applications of this last approach. In the decades before 1800, the German mineralogist Abraham Gottlob Werner created a model of earth history based on the other key "direction" built into Buffon's cosmology, the retreat of a great ancient ocean. But Werner's succession of geological periods was firmly based on empirical evidence: it established a sequence in which the different formations of rock were laid down. He made little use of fossils and incorrectly assumed that each type of rock was only deposited during a single geological period. Nevertheless, the idea that the succession of formations marked a sequence in time, the principle of stratigraphy, became firmly established by Werner and his followers, who were known as "Neptunists" (after the Roman god of the sea).

Like Buffon, Werner did not believe that earth movements were powerful enough to cause major changes to the surface. For this reason his Neptunist theory was routinely criticized by earlier historians of geology. Some of his followers also tried to argue that Noah's flood was a real event—a catastrophic re-emergence of the waters of the ancient ocean—and this was seized upon as evidence that the theory was a conservative effort to resist the development of a more realistic view of earth history (Gillispie, 1959). Later historians of geology have recognized how unfair this characterization is (e.g. Laudan, 1987; Oldroyd, 1996). Most Wernerians did not believe in a recent flood, and imagined a vast sequence of events predating the period in which human life was possible on the earth. Once it was realized that geological periods are best identified by their fossil contents, not by the mineral composition of their rocks, the Neptunist approach became the basis for establishing the modern system of geological periods.

The other development that was needed to create the modern view of earth history was the replacement of the retreating-ocean theory with Hooke's view that earth movements could elevate mountains and continents from beneath the ocean where the sedimentary rocks are laid down. Conventionally, the emergence of this modern viewpoint is associated with the Scottish geologist James Hutton, whose *Theory of the Earth* of 1795 is supposed to have established the true extent of geological time. Hutton was a leading figure in the talented group that made Edinburgh a center of culture in the late eighteenth century—his friends included the philosopher David Hume and the economist Adam Smith. His vision of the earth's history was sustained by the idea of an eternal balance between destruction and creation. He imagined erosion by rain, ice, and rivers gradually wearing away the mountains over vast periods of time, while earthquakes gradually built up new continents elsewhere. Because he thought that earth movements and volcanic activity were produced by a core of intensely hot material at the cen-

ter of the earth, his theory became known as "Vulcanism" after the
Roman god of fire. For Hutton, there had been cycles of the destruction and creation of land stretching over immeasurable periods of time. Although superficially similar to some aspects of the
modern viewpoint, historians now recognize that in some respects
Hutton's was a very un-modern theory. The trouble is that his cycles of erosion and elevation were almost literally eternal—there
could be "no vestige of a beginning, no prospect of an end." Hutton
was a deist, who imagined a God who created a perfectly designed
earth capable of sustaining life indefinitely. He had no interest in
establishing a sequence of geological periods.

The next generation of geologists accepted that earth movements raised new continents, but they were reluctant to evoke the
vast periods of time necessary for modern earthquakes to achieve
such large-scale effects. Instead they supposed that there had been
occasional episodes of violent uplift separated by long intervals of
relative calm. This was the theory of "catastrophism," which has
also received bad press from historians of geology because it limits
the time span of earth history and encourages the possibility that
the last catastrophe might be identified with Noah's flood. The
English geologist William Buckland, an ordained clergyman who
taught at the notoriously conservative University of Oxford, is often
vilified as an exponent of this last-ditch attempt to salvage a scriptural geology. Buckland was a colorful figure who became notorious for trying to eat an example of every known species (including
the rare ones found only in the zoo). But he was a skilled anatomist
and geologist, and an effective lecturer who did much to create an
opening for the new science in conservative religious circles. In
1823 Buckland claimed he had found evidence of a geologically recent flood covering the whole of Europe. At Kirkdale in the York-
shire hills he studied a cave filled with dried mud containing the
bones of species no longer found in England. He was able to show

that the cave had once been a hyenas' den, and argued that the bones here and in similar caves elsewhere must have been buried when a flood swept across the face of the earth.

The deposits Buckland studied were eventually seen as relics of the Ice Ages, and it would be easy to portray him as being led astray by his enthusiasm to re-establish a worldview consistent with Genesis. But his catastrophism was not an effort to defend a young earth (Rupke, 1983). Buckland knew perfectly well that the superficial deposits we now see as products of the ice-sheets were laid down at the very end of a vast sequence of periods in which uplift had alternated with long periods of tranquility. Later catastrophists would routinely accept an age of 100 million years for the earth—well short of modern estimates, but way outside Ussher's chronology. The idea of a diminishing level of geological activity also made good sense if the heat of the earth's core was responsible—how could Hutton's vision of an eternally active earth be squared with the laws of physics, which establish that any hot body must cool down?

The clearest evidence that catastrophism was not a backward-looking approach to earth history derives from the fact that it provided the conceptual foundation upon which the modern identification of the geological periods was based. In England, the canal builder William Smith recognized that the rock strata through which his laborers had to dig could best be identified by the different fossil shells they contained. Evidently, each period of the earth's history had been inhabited by a different set of species adapted to the conditions of the time. In France, the anatomist Georges Cuvier studied the fossil bones being dug up from the various strata and realized the same point—each period has its own characteristic species of animals (Coleman, 1964; Outram, 1984). Cuvier could reconstruct the original form of the animal from the incomplete fossil bones, and he could show quite conclusively that the ancient species were now extinct. Nothing like the mammoth or the mastodon—giant ancient elephants—could be alive today without explorers having discovered it. Yet these were species distinct from the living elephants, so they must now be extinct. Mammals from the older rocks were unlike anything alive today, establishing a general rule that the further back in time one goes, the greater the difference from the modern species.

Lower down in what would soon be called the Mesozoic rocks, there seemed to be no mammals at all, and in 1831 Gideon Mantell coined the term "Age of Reptiles" to denote this period. He had described the teeth of a giant herbivorous reptile he called *Iguanodon*. Buckland had already described a giant carnivore he called *Megalosaurus*. In 1841 the anatomist Richard Owen introduced the term "Dinosauria" to describe these ancient reptiles, and the dinosaurs have remained the best-known examples of the Age of Reptiles ever since. Beneath the rocks of the Age of Reptiles there were formations with no land creatures at all, and the oldest rocks in the fossil-bearing series seemed to contain only invertebrates. The modern sequence of the geological period was beginning to be established, and most of the names we still use for them were coined by catastrophists. Far from hindering the advance of geology, catastrophism encouraged the recognition that there were distinct periods of earth history, even if establishing the exact boundaries was a complex business requiring much negotiation and sometimes acrimony among the experts (Rudwick, 1985). The catastrophes could be seen as convenient punctuation marks dividing geological time into recognizable units. Modern explanations of mass extinctions have reintroduced a role for catastrophes, as with the widely accepted view that the impact of an asteroid was responsible for wiping out the last of the dinosaurs.

The geologists of the mid-nineteenth century had thus developed a model of the history of life that was remarkably similar to the modern one, with a single crucial difference. They believed that species became extinct suddenly, wiped out by violent earthquakes and tidal waves that devastated whole continents, if not the whole earth. And when the slate was thus wiped clean of life, they assumed that God would step in and create a new population of species adapted to the next interval of stability. Paley's argument from design could be preserved by imagining not a single creation in the Garden of Eden, but a sequence of such events occurring at the beginning of each geological period. Buckland wrote a book in a series commissioned by the Earl of Bridgewater to expand the claim that science revealed the "power, wisdom and goodness of God." His *Bridgewater Treatise* (1836) surveyed the whole known fossil record, reconstructing what the extinct animals had looked like and arguing that each species was well-adapted to the environment of the period in which it lived. If the species became step-by-step more similar to those we know today, that was because the earth's environment was gradually becoming more benign as the planet cooled down. God could thus create successively higher forms of life in each geological epoch, culminating with the appearance of humans (Bowler, 1976a, Rupke, 1983). Buckland certainly did not believe that the physical catastrophes were supernaturally caused— the cooling-earth theory explained why past events were more violent.

Buckland's vision was still bounded by the idea of divine creation, even if extended vastly beyond the traditional understanding of the biblical story. Most educated people of his time were not deeply concerned by the apparent undermining of Ussher's timescale of earth history. Biblical scholars were already beginning to realize that the books of the Bible could be understood as historical records, whatever the spiritual message they conveyed. They were limited by the conceptual system of the people who wrote them, and thus spoke in allegorical terms when dealing with events outside human experience. The creation story did not have to be taken absolutely literally, as long as its spiritual message—that the earth is a divine creation—is respected. There is, in fact, no evidence that the early saints and scholars who formulated the basis of Christian teaching took the Genesis creation story to mean that the earth was created in seven 24-hour days. It was the Protestant theologians of Ussher's time who focused their attention so closely on the written text that they felt compelled to take it all literally.

Two different approaches were available to those looking for a less restrictive understanding of the Genesis story. As Buffon had somewhat cynically noted, one could understand the term "day" to refer to a time period of indefinite length. Thus the main geological epochs became the days of creation. This view was endorsed in the nineteenth century by geologists such as Hugh Miller and J. W. Dawson. Alternatively, the text of Genesis seems to allow for a gap between the creation of the universe and the events in the Garden of Eden. If this gap were an enormous period of time, it could contain all the events of earth history not actually recorded in the text. This interpretation was favored by William Buckland and Adam Sedgwick, and was promoted in America by Edward Hitchcock. Both approaches allowed the catastrophist geologists to accept a vastly extended history for the earth while retaining Ussher's timescale for human history. It was an article of faith that no human remains were found in the fossil record, so the human species was a very recent creation. Only in the 1860s did archaeologists' discovery of stone-age tools threatened this compromise position.

The catastrophists' reliance on miraculous creation provides another reason why their theory has been widely, if unfairly, dismissed as a hindrance to the development of the modern form of geology. Historians' attention switched to the emergence of the alternative "uniformitarian" approach, which had been pioneered by James Hutton but was thrust to prominence in the 1830s by Charles Lyell (Wilson, 1972). Lyell had been trained in the law, but had developed an interest in geology through contact with Buckland. He soon became dissatisfied with Buckland's explicit efforts to identify the last geological catastrophe with Noah's flood. Lyell's liberal po-

litical views made him suspicious of efforts to defend the new science by using it to uphold established religion. His *Principles of
Geology* of 1830–33 argued that it was unscientific to postulate catastrophes outside the range of anything experienced by humans.
Only observable causes could be used in a scientific theory, and this
meant that the uplift of mountains was due to the accumulated
effect of normal earthquakes over millions of years. The carving
out of valleys was similarly the result of normal processes of erosion by rain and streams over a vast period of time. For Lyell, time
replaced violence in the explanation of how the earth's surface was
shaped. The hundred million years accepted by most catastrophists
for the total age of the earth was nowhere near enough to have produced the effects we observe in the geological record. Lyell probably
thought in terms of billions of years—correctly according to modern estimates—although he would not name an actual figure. In
fact Lyell shared Hutton's view that the cycle of creation and destruction had gone on more or less eternally. In Stephen Jay Gould's
view (1987), Hutton and Lyell adopted a steady-state view of earth
history, not one based on a genuine historical sequence in which
the past was significantly different from the present.

Lyell's theory was effective not so much in converting the catastrophists (it didn't) but in convincing everyone that modern
causes were active enough to have produced at least some of the
changes visible in the geological record. His greatest disciple was
Charles Darwin, who saw evidence while on the voyage of HMS
Beagle that the Andes mountains of South America have been
raised gradually by earthquakes similar to those which still affect
the region. Darwin went on to extend the uniformitarian method
into an area that Lyell avoided, the origin of new species. Lyell accepted that extinction must be a gradual process occurring as conditions gradually changed and became less suitable to the existing
species. But he would not accept that natural causes could actually
change species to adapt them to the new environment. Lyell still be-

lieved in divine creation, although for him species would have to be created from time to time on a more or less continuous basis.

CHALLENGING DESIGN

The catastrophist position was quite modern in some respects, yet it stopped short of the last major step that was needed to formulate an evolutionary viewpoint. It accepted that the earth's physical history was a process governed by the normal laws of nature, such as the law of cooling. But on the question of how new developments took place in the world of life, many catastrophists remained committed to the idea of miraculous creation, or at least of some supernatural involvement which allowed new species to be seen as products of divine wisdom. Nature could sustain life, and even destroy it on a wholesale basis, but natural processes were incapable of any truly creative act.

From the start there were some skeptics who were suspicious of the desire to retain so close a link with the traditional model of divine creation. The eighteenth century became known as the Age of Enlightenment, as European thinkers became convinced that human reason could understand the laws governing the world, life, and even society (Israel, 2001; Outram, 1995; Porter, 1990). There was a new air of confidence, generating a willingness to challenge traditional authorities when they seemed to block the advance of rational investigation and reform. In the end, this attitude would lead to the political upheavals of the American and French revolutions, with far-reaching consequences for the subsequent development of Western societies.

More directly relevant to our present topic is the willingness of some more radical thinkers to challenge the Christian religion in general, and the biblical story of creation in particular. From these challenges came some of the first naturalistic alternatives to the traditional vision of divine creation. These early speculations often bear no resemblance to the modern theory of evolution, which is why I will use the contemporary term "transformism" to describe them. Far from treating their authors as "forerunners of Darwin" (as in Glass, Temkin, and Strauss, 1959), we should see them as evidence of just how difficult it was to formulate the modern theory of evolution. Darwin's model, based on the divergent evolution of related species from a common ancestor, was not a self-evident solution to the problem. Radical naturalists explored a number of quite different alternatives, which have been misunderstood by later biologists and historians. Too much hindsight has been applied to these theories, making them seem like precursors of Darwinism when in fact they were pointing in very different directions.

As radical naturalists began to explore different ways of understanding how the succession of living things might be produced, conservatives were forced to rethink the idea of design. Some were eminent scientists in their own right, and they did not want to be dismissed as out of touch with the latest developments. Even the opponents of transformism had to accommodate themselves to the possibility that the creation of living things followed law-like trends which looked superficially very similar to what the transformists were predicting. In the decades before Darwin published the *Origin of Species*, the idea that the history of life might unfold by law rather than by a succession of miracles began to gain some degree of credibility. In one sense this paved the way for the reception of Darwin's theory, but it also shaped—and perhaps distorted—the way in which the public perceived the message contained in his book.

One figure who has routinely been misidentified as a forerunner of Darwinism is the comte de Buffon, whose estimate of the earth's antiquity challenged the literal reading of Genesis in 1749 (translations in Buffon, 1981; see Bowler, 1973; Roger, 1997). Buffon was a follower of Newton, who hoped to extend the realm of naturalistic explanation to the whole physical and biological world. His suggestion that the earth was formed when a comet struck molten material from the sun was the first step in this campaign. But when it came to explaining the origin and development of life, Buffon was far less decisive. As he wrote the successive volumes of his *Natural History* he circled around the problem, sometimes contradicting himself. He argued that the relationships between species used by Linnaeus to classify them into groups had to be based on something real—he had no time for the assumption that the relationships were part of God's great plan. Yet he could not shake off the traditional belief that species were real and fixed, as permanent as the laws of nature. In the end, he accepted that the closely related species linked by Linnaeus into a single genus were the products of a natural process of transformation in the course of time. The lion, tiger, and leopard had diverged from a common big-cat ancestor, perhaps as the original population was split up by migration to different parts of the world. Yet at the same time he declared that the modern forms were not true species. The lion, tiger, and leopard were merely very strongly marked local varieties of a single big-cat species. If Buffon took one step toward the modern idea of divergent evolution, he took another step back to the traditional assumption that the main types of life are permanently fixed.

But where did the ancestors of the big cats and the other main animal types come from? As a materialist, Buffon was certainly not going to concede that they were created by miracles at certain points in the earth's history. Instead he appealed to the ancient notion of "spontaneous generation," according to which maggots, for instance, were produced naturally from decomposing meat. Francesco Redi had already shown that this explanation was wrong in the case of maggots—they hatched from flies' eggs laid on the meat. But Buffon performed experiments which seemed to show that micro-organisms, at least, could be produced by such a spontaneous coming together of the "organic particles" in meat juices. In fact, he had not sterilized the apparatus properly, as Lazarro

Spallanzani showed, but this took some time to establish with complete certainty. In the meantime, Buffon could point to his experiments as evidence that unaided nature could spontaneously assemble living things from a loose collection of "organic particles." If micro-organisms could be produced in the laboratory, who knows what nature might generate in the vast, warm oceans of the ancient earth? In his *Epochs of Nature* of 1778, Buffon postulated two major episodes of spontaneous generation, the first forming the fossil species that eventually became extinct, the second producing the ancestors of the major animal types of today. He hinted that exactly the same species would be generated on the other planets of the solar system when they reached the right temperature.

Some of the radical French philosophers with whom Buffon was on good terms were even more aggressive (Roger, 1998). Denis Diderot was the editor of a great *Encyclopedia* that criticized many of the assumptions on which the old social order was based, paving the way for the French Revolution of 1789. He was an atheist who went much further than Buffon in attacking the religious beliefs upon which the Church based its authority to uphold the divine right of the king and the aristocracy. In works which were so subversive that they had to be circulated in secret (modern translation in Diderot, 1966), he imagined a nature that was subject to constant flux, with no fixed species—and hence, by implication, no fixed social classes. There was no divine plan. Buffon was right to imagine episodes of spontaneous generation, but these did not produce predetermined types that might be mistaken for divinely ordained units in a plan of creation. No, nature just produced living things at random, their organs pasted together without plan or purpose. Most died off immediately, but a few by chance could survive and breed, becoming the ancestors of the species we know today. Even then, there was nothing to guarantee the fixity of the type, and the occasional appearance of monstrosities in the course of normal reproduction suggests that the process of trial and error is still going on. At one level, Diderot anticipated the logic of natural selection, but his was no scientific theory of gradual evolution. It still imagined that complex animals could be formed directly by spontaneous generation. And the hint that monstrosities might be a source of new species offered no explanation of how they would perpetuate themselves. Diderot's ideas were a speculative assault on the logic of the argument from design, not a scientific theory of evolution.

Diderot's rejection of design was paralleled in the views of the Scottish philosopher David Hume, whose *Dialogues on Natural Religion* also attacked the logic of the analogy between living species and human artifacts. We can see watchmakers building watches, but no one has ever seen a species produced by a miracle, so how do we know that the same element of design is involved? Nature looks more like a gigantic living organism than a colossal machine, so any application of a model based on engineering is invalid. Hume offered no rival theory to explain how species became adapted to their environment, but some historians think that his philosophy of overall skepticism was an important influence on Darwin. Hume denied that we can see the underlying causes that operate in nature—all we observe are regularities in the phenomena we investigate. Those who claim to see clear evidence that there is a divine plan are thus deluding themselves. We need to look more carefully at what regularities actually occur in nature—and as Darwin found, the closer you look, the less secure the idea of the absolute stability of species seems to be.

THE FIRST TRANSFORMISTS

The Enlightenment skeptics challenged the argument from design, but they can hardly have been said to have formulated a workable alternative that naturalists could use to explain the development of life. To some extent, the revival of the ancient notion of spontaneous generation served as a distraction, deflecting attention away from what we now see as the more plausible approach based on the transmutation of existing species. The eighteenth-century materialists were more interested in the origin of life—speculating about its variability once formed presented less of a challenge to the model of divine creation. But the claim that unaided nature might have generated complex animals from disorganized matter, even at some remote point in the past, was beginning to seem less plausible. Buffon's experiments did not stand the test of time, as Lazzaro Spallanzani demonstrated when he showed that if the meat juices were properly sterilized, no micro-organisms were produced. Those favoring materialism continued to believe that perhaps the most basic form of life might be produced directly by nature, and by the end of the century the discovery of the electric current offered them a new force which, it was hoped, might explain how dead matter could be brought to life. But the process of spontaneous generation could only start the ball rolling for the development of life. It could produce the most basic starting point, but more complex creatures would have to be developed in the course of time from these primitive origins. Life would have to change through time, making transformism an essential component in the skeptics' assault on the design argument. Species would have to become steadily more complex, introducing the idea of progress into the organic world. And they would have to find ways of surviving in the conditions they encountered, making adaptation a natural process rather than a divinely preordained state.

Any suggestion that living species, including the human species, could be produced by natural means threatened the traditional interpretation of Genesis. The emergence of these early ideas cannot be understood except in the context of a radical intellectual movement that favored skepticism, materialism, and even atheism. But these were also ideas with social implications: the Church served as a bastion of the traditional social order based on rule by a monarch and an aristocracy. The social hierarchy was divinely established, so to challenge it by calling for reform was to challenge the way God had ordered the universe. Transformism undermined the credibility of Genesis, thereby threatening the claim that there was a divinely preordained structure built into the world. If nature could change, then why not society? Social thinkers such as the marquis de Condorcet appealed to the idea of progress to justify calls for reform. They argued that throughout history, society had been gradually changing as humanity developed better ways of exploiting nature. If progress had occurred from the earliest times to the present, then it was only natural to expect that it would continue into the future. Future progress, of course, meant reform of the existing social order. Expanding the idea of progress into the organic world merely extended this argument—progress was inherent in both nature and society. The improvement of society was an inevitable outcome of processes built into the very fabric of the universe. Many who favored reform welcomed the idea of transformism once it was linked to the ideology of progress (Ruse, 1996).

Diderot's *Encyclopedia* was meant to show how new trades were transforming society. By implication, those who understood and practiced those trades had a better right to govern than the aristocracy for whom land was the only source of prestige and power. Such ideas would help to destabilize the French monarchy and lead to the Revolution of 1789. Here the demand of the middle classes for power and influence would be worked out in blood, both on the guillotine and on the battlefields of Europe. But in Britain the middle classes had a head start and were able to engineer a transformation of society which, despite various episodes of confrontation, was gradual and relatively peaceful. The inventors and entrepreneurs who built the Industrial Revolution became rich as the economy expanded and were able to use their wealth to buy social position and political influence. They were anxious to use the latest developments in science both to promote technological innovation and to bolster the ideology of free enterprise on which their success was based. Here too the idea that nature was an ever-changing system of progress and adaptation provided the basis for an alternative to the conservatives' insistence on stability and the divine order. The radicals who represented the laborers who worked in the new industries wanted an even greater level of reform that would benefit the whole of society, not just the managers. Those who had everything to gain from social change welcomed the progressionist model of transformism with open arms.

Two figures are widely identified with the introduction of a completely transformist vision of how nature develops. In Britain, Erasmus Darwin is remembered both for his own pioneering contributions, and for the fact that it was his grandson, Charles Darwin, who conceived and popularized the modern version of evolutionism. In France, Jean-Baptiste Lamarck published more extensive accounts of a similar view of natural development and founded a movement that preceded and for a while seemed to offer a viable alternative to Darwinism. Erasmus Darwin was a member of the new middle-class elite that was bidding for power and influence in the increasingly industrialized society of Britain. Lamarck worked at the restructured Museum of Natural History created by the revolutionary government in France. Both linked spontaneous generation with the idea that living things have progressed and adapted over vast periods of time to generate the variety of forms we see around us today.

A successful physician, Erasmus Darwin founded the family fortune that would one day sustain his grandson in comfort (Browne, 1989; McNeil, 1987; Porter, 1989). He worked not in London but in the English Midlands, the heartland of the Industrial Revolution. He was a close friend of James Watt and Matthew Boulton, who were introducing steam power to drive the machinery of the new industries, and of Josiah Wedgwood, founder of the pottery firm that was one of the great success stories of the age. These were all people who appreciated the power of innovation to transform the economy and society itself. They were mostly nonconformists in religion, opposed to the Church of England's function as a bastion of the old social order. Some, like Erasmus Darwin himself, were inclined to a deism that represented God as the Creator of a self-developing universe. He saw how the ideology of progress could be translated into the organic world to give a vision of universal history that would serve as a foundation for the model of social progress. He was also a poet, who articulated this image in verses that became quite popular in the era just before the Romantic movement transformed literary tastes.

As a medical man, the older Darwin was certainly aware of the ideas proposed earlier by radical thinkers such as Buffon. He was prepared to accept that unaided nature could generate simple living organisms: God did not need to create by miracle, because the laws He had established were designed to allow natural structures to increase their level of complexity. Once life had appeared, each organism strove to adapt itself to its environment and to reproduce. Some forms became predators, and Darwin was well aware of the role played by struggle in nature. Some historians have tried to see in this aspect of his thought an anticipation of his grandson's theory of natural selection. But recognizing the prevalence of bloodshed and cruelty is not the same as seeing the selection of random variation as a creative force. Erasmus Darwin was more interested in how individuals improve themselves through their own efforts. He was fascinated by the process of reproduction (and not just intellectually, since he fathered a large family of children). Like Lamarck, with whom the idea is normally associated, he simply assumed that any characters acquired by the organism in the course of its efforts to survive would be reflected in the process of reproduction and thus passed on to the offspring. Thus as individuals improved themselves, so did the race, because each generation was born with the accumulated benefits acquired by its forebears. Darwin articulated this idea into a vision of universal progress in which life ascended the scale of complexity—usually known as the "chain of being"—through its own unaided efforts at self-improvement. Humans were merely the highest products of this universal trend, and modern social progress was its inevitable continuation.

Erasmus Darwin's most systematic account of his theory came in a single chapter of his book *Zoonomia: or the Laws of Organic Life* (1794–96). But because this was a physician's vision of life, it paid little attention to the problems of a working naturalist trying to understand the relationships among species and their succession in space and time. His impact came more from his poetic vision, which occasionally broke out into an ecstatic vision of progress, as in *The Temple of Nature* of 1803 (lines 295–302):

> ORGANIC LIFE beneath the shoreless waves
> Was born and raise'd in Ocean's pearly caves.
> First forms minute, unseen by spheric glass,
> Move on the mud, or pierce the watery mass;
> These, as successive generations bloom,
> New powers acquire, and larger limbs assume;
> Whence countless groups of vegetation spring,
> And breathing realms of fin, and feet and wing.

These lines would have been read in many a polite household, and would certainly have come as something of a shock to anyone used to thinking in terms of divine creation. Paley's *Natural Theology* was in part a response to the challenge posed by Erasmus Darwin's transformism.

Jean-Baptiste Lamarck's theory was a more effective assault on the old, static vision of nature because it was disseminated from the heart of the new scientific community established in revolutionary France. Lamarck was an eminent naturalist who made major contributions to the classification of the invertebrates. His theory may have been speculative, but it could not be dismissed as the product of an imagination out of touch with the realities of nature. By modern standards, the theory is indeed a bizarre framework which articulates the idea of natural development in a way that is very difficult for us to comprehend properly. Historians are now convinced that it is a mistake to treat Lamarck as a forerunner or precursor of Darwin (Barthélemy-Madaule, 1982; Burkhardt, 1977; Corsi, 1988a; Hodge, 1971; Jordanova, 1984). Yet the impact of this rival to natural theology would reverberate among the radical thinkers of the early nineteenth century in a way that would certainly shape the world in which Charles Darwin functioned. The conservative establishment might try to dismiss Lamarck as a crazed visionary, but radicals looked to his work for ammunition to use in their campaign to discredit the ideology of a hierarchical world established by divine fiat.

Lamarck was originally a botanist, and at one time served as the tutor to Buffon's son. When the new revolutionary government in France reorganized the country's scientific institutions and created a Museum of Natural History in Paris, Lamarck was appointed in 1794 to work on the invertebrates. The more prestigious task of describing and classifying the vertebrates was given to the man who would eventually try to discredit him, Georges Cuvier. Whereas Cuvier described the vast array of fossil bones being unearthed in terms of fixed species that ultimately became extinct, Lamarck began to question the stability of living forms and proposed his theory of transformism. This was already being formulated around 1800, but its best-known expression is the *Zoological Philosophy* of 1809 (translated as Lamarck, 1914).

Like Erasmus Darwin, Lamarck assumed that the simplest forms of life could be produced by spontaneous generation. He appealed to one of the great discoveries of the age, the electric current, which Luigi Galvani had shown was intimately connected with the processes of life. A dead frog's legs could be made to jump when an electric shock was passed through them. Significantly, in terms of the popular imagination, Mary Shelley's gothic novel *Frankenstein,* written in 1818, imagines the monster being brought to life by electricity. Lamarck knew that unaided nature could not produce complex living structures, but he imagined that the electric fluid could generate the simplest forms of life. He was a uniformitarian in geology, convinced that the earth had undergone vast cycles of gradual change through time. The earliest forms of simple living organisms had been produced in the distant past by spontaneous generation. Gradual transformations over untold generations had advanced these earliest forms up the scale of complexity to become the higher animals of today. But here we encounter one of the crucial differences between Lamarck's theory and modern evolutionism. Charles Darwin's theory assumed that all later forms of life, high and low, are the descendants of those early, very simple organisms. Lamarck thought that the spontaneous generation of simple living organisms had gone on throughout the earth's history and is still at work today. He posited that the reason why we still have simple forms of life in the world is because they have recently been produced by fresh acts of spontaneous generation.

Lamarck imagined that evolution is like an escalator transporting successive generations steadily up the scale of development. The problem with this assumption is that if you imagine life always becomes more complex, how can there be simple forms left in the world? The only way out is to argue, as Lamarck did, that the simple forms at the bottom of the scale are constantly being renewed. Lines of progressive development are constantly setting off up the escalator, all running up the same hierarchy of forms in parallel, one behind the other. The highest forms of today, presumably ourselves, are the product of the oldest line of development and have had the longest time to mount up the scale. This is a profoundly non-Darwinian way of visualizing organic change. It denies the branching nature of evolution on which the classification of species into groups of forms related by common descent is based. As we shall see in later chapters, the model of parallel developments ascending the same predetermined scale served as one of the most potent alternatives to Darwinism and was only finally discredited in the early twentieth century. Those who casually assert that evolution is based on the idea of inevitable progress should remember this point.

Two further points reveal how far Lamarck was from our modern viewpoint. First, he did not believe that there were distinct species in nature. He was convinced that natural forces operated so as to produce a complete continuity of forms: there were no sudden jumps in development, so there could be no distinct species. This is the basis of a common argument still used against Darwin: If evolution is a gradual process, where are the intermediates between the apparently distinct forms we observe? This is a misunderstanding derived from a failure to realize that evolution is a branching, not a linear process. There are gaps because the lines of development are split from time to time (by geographical barriers, for instance) and the separate populations then move off in their own distinct directions. As they move away from the common ancestor, gaps open up between them that allow them to be recognized as distinct species.

The second difference between Lamarck's view and modern evolutionism is that he did not accept the possibility of extinction. Exponents of natural theology had always been reluctant to concede that God would create species only to allow them to disappear in the course of time. Lamarck too felt that nature's powers were so great that none of her products could disappear without issue. When Cuvier reconstructed ancient species from their fossil remains and claimed that they were no longer in existence, Lamarck insisted that they had simply changed into something else which is still alive in the modern world. The idea that evolution is constantly branching, so that some species die out while others divide into multiple descendants, did not occur to him.

We now have to assess the one part of Lamarck's theory which—

although incorrect according to modern genetics—was neverthe-less to play a significant role in the history of evolutionism. Al-though he thought that there was a progressive force driving all forms of life in parallel up the scale of complexity, he was an experi-enced naturalist who accepted the point stressed by Ray and Paley in their version of the argument from design. All species seem well adapted to their way of life, a situation that the exponents of natu-ral theology took as evidence of divine benevolence. Lamarck sug-gested instead that there was a natural process which could gradu-ally change living things so that they adapted to the environment in which they had to live.

Charles Darwin would later postulate what he called natural se-lection to explain how adaptive structures are built up. But Lamarck appealed instead to an old idea, the inheritance of acquired charac-teristics. Everyone knows that animals can acquire new characters if they change their behavior to exercise their bodies in a new way. An obvious illustration of this in human affairs is the bulging mus-cles of the weightlifter, acquired thanks to many hours of pumping iron. Animals too will acquire new bodily characters if they change their habits. But what happens when they subsequently reproduce? Will the acquired character be inherited, even to a slight degree, by their offspring? Will the weightlifter's children grow up with slightly larger muscles than they would have had if their father (and, for a better effect, their mother too) had not built their mus-cles pumping iron? Many people assumed that there is a tendency for the acquired characteristics to be passed on, and Lamarck also believed this would happen. If one thinks of heredity as a process by which the whole organism transmits its characters to its off-spring, then this assumption is perfectly natural. Unfortunately, modern genetics tells us that this is NOT the way heredity works: only those characters already coded in the genes can be transmitted, and effects developed by the body in addition to its genetic inheri-tance are irrelevant to the reproductive process.

In Lamarck's day, however, genetics lay a century ahead and there was little to challenge the popular assumption that acquired characters can be transmitted. Lamarck realized that such a process, acting over many generations, could explain adaptive transmutation. If animals modify their habits to cope with a changed environment, they will acquire new characters adapting them to the new habits. And if those characters are transmitted, the next generation will be born with that character already slightly better developed, and will add to it thanks to their own continued efforts. Over many generations the new character will gradually be built up until the species is perfectly adapted to the environment. The classic example (although Lamarck himself only mentioned it briefly) is that of the giraffe's long neck. Assume that the ancestors of the modern giraffes were mammals with normal necks something like a modern antelope. If they were confronted with a changed environment in which the leaves of trees were the best source of food (perhaps because the grass was disappearing), they would exercise their bodies in a new way by stretching up to reach the branches of the trees. Their necks would become slightly elongated—and their offspring would be born with longer necks, which they would stretch further as they too continued the new feeding habit. Over many generations of consistent exercise, the neck would become massively enlarged, becoming the giraffe we know today. The neck of the modern giraffe is, in this example, the product of its ancestors' efforts accumulated over many generations.

In the later nineteenth century, many naturalists would accept the inheritance of acquired characters as an explanation of adaptive transformation. Lamarck was remembered as one of the first to suggest the mechanism, and so it became known as "Lamarckism." Even Darwin thought the effect was real, although he subordinated it to his own idea of natural selection. Only with the advent of Mendelian genetics in the twentieth century did its plausibility diminish. But we must be careful when we evaluate the link between

Lamarck himself and the later Lamarckians. They took the one element of adaptive evolution from Lamarck's theory and applied it in a very different way. For Lamarck himself, the inheritance of acquired characters was only a secondary process, modifying the force of progressive development that pushed living things up the scale of complexity. He did not explore the idea that related species have diverged from a common ancestor, and—as already noted—he rejected the possibility of extinction. His was certainly a complete theory of transformism, but it was built on different foundations to those which Darwin would employ. The difference between Lamarck's theory and Darwin's is far more substantial than the mere replacement of the inheritance of acquired characters with natural selection.

Until comparatively recently, most historians assumed that there was, in fact, no real continuity between Lamarck's work and that of later generations of evolutionists. In his own time, his ideas were rejected by the conservative scientific establishment in both France and Britain. It was the politically astute Georges Cuvier who emerged as Lamarck's most severe critic, effectively marginalizing him as a visionary and a crank. The whole idea of transformism was driven underground, to re-emerge only when Darwin reconfigured the theory along modern lines. But thanks to the work of historians such as Toby Appel (1987), Pietro Corsi (1988a), and Adrian Desmond (1989), we now know that there is a more complex story to be told about the intervening period. Lamarck was indeed rejected by the conservatives, but his ideas were taken seriously by the radical thinkers who circled around the elite scientific community. These were people who were often both politically as well as intellectually radical—they wanted to explore new ideas that seemed to threaten the foundations of the old worldview. If transformism could be used to undermine the credibility of the static model of creation, they would exploit it. The conservatives pretended that they were ignoring the challenge, but in fact they were looking over their shoulders at a baying mob of critics.

In France, another rival of Cuvier—Étienne Geoffroy Saint-Hilaire—adopted a rather different explanation of transformism (Appel, 1987). He supposed that massive disruptions of embryological development might produce individuals with characters sufficiently different to be classified as members of a new species. This is transformation by sudden leaps or saltations, parodied in later times as the theory of the "hopeful monster." Geoffroy's was a materialistic science that sought to explain the underlying similarities of related species in naturalistic terms. In Germany, there was a similar focus on the underlying unity of nature, as an alternative to the utilitarians' vision of each species being designed as a unique adaptation. In Britain, Lamarck's own ideas were synthesized with this "transcendental anatomy" by a generation of radical anatomists who challenged the elite of the medical profession. They used the various ideas of transformism to undermine the static model of creation preferred by the established figures in the Royal College of Surgeons (Desmond, 1989). Many were political radicals too, if not outright revolutionaries. Transformism acquired the reputation of being a materialistic science because it was being used by those who sought to challenge the idea of design. By ridiculing the science of the elite as out of date, these radicals hoped to undermine the static model of creation and free things up so that reform could take place.

One of the few exponents of this radical science remembered in conventional histories is Robert Edmond Grant. Darwin met Grant while he was a medical student in Edinburgh in the 1820s and later recalled how shocked he was to hear Grant expound his transformist views. Historians now suggest that Grant may have had a much greater influence on the young Darwin than the latter was later prepared to admit (Sloan, 1985). Darwin may not have taken Lamarck's theory on board, but he became aware that there were people looking for underlying unities in nature which could only be explained in terms of transmutation. Although he went off to Cambridge and absorbed a more conservative worldview based on Paley, when he eventually reconsidered the case for transformism, the memory of Grant's Lamarckism returned to warn him of how dangerous such ideas could be.

REINTERPRETING DESIGN

Grant and his radical allies represented the main threat against which Christian writers declaimed. For the radicals, the new science offered the hope of replacing religion with a completely materialistic worldview. But there were many who wanted to preserve traditional religion while allowing the new science to flourish. In the established church, natural theology was still highly valued, and a new generation of devout scientists hoped that it could be reformed in a way that would accommodate the latest insights without undermining the argument from design. It has to be said, however, that some of their efforts ended up looking very similar to the ideas of the secularists. This was the trend that many evangelicals feared, and although some of the more liberal evangelicals took on board aspects of the new theories of development, many of the science writers from this background focused on traditional aspects of natural history that did not raise the more threatening issues.

The social influence of religious groups was still considerable. Grant himself was forced into obscurity when he left the safe haven of Edinburgh to take up a position at the newly founded University College in London. This was an institution founded by nonconformists (because Oxford and Cambridge still admitted only Anglicans), and it was more open to new ideas. But Grant's radicalism went too far, and although he kept his position he was gradually marginalized within the scientific community of the capital. The man who engineered his downfall was the rising star of British comparative anatomy, Richard Owen (Desmond, 1982; Rupke, 1994). As one of Darwin's chief opponents following the publication of the *Origin of Species,* Owen has been vilified as an arch conservative in conventional histories of evolutionism. He was certainly a difficult figure who alienated many of his fellow anatomists—but whether or not their harsh judgments can be taken at face value is open to question. It is worth remembering that the Natural History Museum in London owes its present form to Owen. (If you go to the Museum today you will see a statue of Owen at the head of the main staircase, whereas Darwin and Huxley are consigned to the tea-room beneath his feet.) Owen made his reputation among the conservative scientific elite by defending the idea of design—but by the standards of the time he was an advanced thinker who brought new ideas into comparative anatomy.

To understand the paradoxes of Owen's position—and the complex social environment within which the early ideas of transformism had to operate—we need to appreciate that there are many different ways of thinking about organic relationships. Owen knew that comparative anatomy was being stifled by the conventional view that each species should be treated as a unique case of adaptation. The whole point of comparing the internal structures of species was to show how they were related to one another. The Germans had shown that one could visualize these relationships as different adaptive modifications of an underlying basic pattern. In his *Archetype and Homologies of the Vertebrate Skeleton* of 1848, Owen introduced this Continental transcendental anatomy to Britain and presented it as an updated version of the argument from design. The vertebrate archetype was the idealized basic form of all backboned animals, of which the major classes (mammals, birds, reptiles, fishes) are fundamental modifications. The same kind of relationship could be seen at a finer level: the human hand, the wing of the bat, the flipper of the whale, all showed the same underlying pattern of bones modified for different adaptive purposes. For Owen, these "homologies" offer a better indication of design than the adaptations themselves—they show that the Creator is a rational God who works with a coherent plan.

Owen offered this interpretation of anatomical relationships as an alternative to the radicals' transformism. Relationships that they (and modern evolutionists) explained in terms of natural transformations, Owen presented as an idealized pattern in the mind of the Creator. Note, however, that the emphasis has shifted a long way from Lamarck's theory—Owen's new vision is fully compatible with the modern system, which pictures relationships in branching rather than linear terms. All Darwin had to do was turn the imaginary archetype into a real common ancestor from which the related species had diverged by adaptive evolution. In the 1820s and 1830s it was vital for Owen to insist that no such natural transformations were possible, and he stressed the distinct character of the various modifications. This included in particular the distinction between humans and their closest anatomical relatives, the apes.

At this point Owen was a confirmed opponent of Grant's transformism. Yet by the later 1840s Owen himself was beginning to realize that one could imagine the adaptive modifications unfolding through time as part of a divine plan programmed into the laws of nature. He was even willing to interpret the succession of species in the fossil record as a progression toward the human form:

The archetypical idea was manifested in the flesh under divers such modifications, upon this planet, long prior to the existence of those animal species that actually exemplify it. To what natural laws or secondary causes the orderly succession and progression may have been committed we are as yet ignorant. But if, without derogation of the Divine power, we may conceive the existence of such ministers, and personify them by the term 'Nature,' we learn from the past history of our globe that she has advanced with slow and stately steps, guided by the archetypical light, amidst the wreck of worlds, from the first embodiment of the Vertebrate idea under its Ichthic [fish-like] vestment, until it became arrayed in the glorious garb of the human form. (Owen, 1849, p. 89)

Note that Owen explicitly mentions natural laws governing the appearance of new forms, not miracles, although he supposes these laws to embody God's design. In the 1850s he would also show how the fossil record could be seen as a sequence of increasingly specialized branches diverging from a more generalized common ancestor—something Darwin would take as evidence of adaptive evolution. Owen would never accept a purely natural form of evolutionism, but his later opposition to Darwin was founded not on a total rejection of common descent, but on a distrust of the materialistic nature of Darwin's explanation of the process. Owen's recognition of the significance of homologies was a genuine advance in biological thinking that paved the way for general acceptance of what we see as the branching model of evolution. Far from holding back the advance of science, a "modernized" version of the design argument was able to generate insights that were of permanent value and could be incorporated smoothly into the Darwinian scheme.

THE VESTIGES OF CREATION

Owen realized that these views would not find favor with his conservative backers. He would only develop his theistic model of evolution after Darwin had published. In the 1830s his opposition to Grant's radical transformism identified him as a key figure in the defense of a discontinuous model of nature's development. New species may come in according to a lawlike pattern, but they are still created separately. No one hoping to enter the elite ranks of the scientific community could afford to be identified with the materialist doctrine of transmutation. Darwin himself, having conceived his theory of natural selection in the late 1830s, was keeping it very much to himself. But as Owen's new ideas showed, the trend in science was to extend the role of law in the explanation of origins at the expense of the more unpredictable elements implicit in the traditional image of divine miracle. Would it be possible to convince ordinary, deeply religious people that God might build laws into his world that would allow His creation to unfold as He intended, but without His continued miraculous interference? In effect, could transformism be freed from the materialist label pinned onto it by Grant and the radicals and presented in a new guise that could be accepted by the middle classes? After all, the respectable leaders of commerce and industry were just as wedded to an ideology of reform which demanded that society be seen as progressive rather than static. What was needed was a popular account of transformism that would link it to the idea of gradual (but not revolutionary) progress.

Such a popular account appeared anonymously in 1844 under the title *Vestiges of the Natural History of Creation* (modern edition in Chambers 1994; associated texts in Lynch, 2000). Its author was the Edinburgh publisher Robert Chambers, although this would not be known for some time, leaving room for endless speculations. Chambers came from just the background one would expect for a proponent of a science of progress, the upwardly mobile middle class. He was not very religious himself, and evangelicals regarded his publishing house as a dangerous influence because its products seldom mentioned spiritual matters. But Chambers knew that to make transformism acceptable at this level, it would have to be purged of its radical image and presented in terms which made progress seem part of a divine plan. His book thus appeared, ostensibly at least, as a contribution to the argument from design: new species appeared in accordance with divinely instituted laws in a sequence determined by the Creator—but the sequence was built in from the beginning and did not need His guiding hand at each step in the advance. Chambers was not a scientist, although he had an amateur interest in natural history, but he argued that the professionals like Buckland and Owen couldn't see the wood for the trees. If they stepped back and looked at the overall history of life on earth, they would see that it unfolded in a more or less continuous trend that was compatible with transmutation rather than a sequence of arbitrary miracles.

For Chambers, the whole history of the earth unfolded according to law. The planets themselves were formed by the condensation of a vast, rotating dust cloud under the law of gravity. This was the "nebular hypothesis" proposed by the French astronomer Pierre-Simon Laplace, and it played a major role in the thinking of those who saw even the physical universe as having an evolutionary history (Numbers, 1977). But *Vestiges* soon moved on to the history of life, and here Chambers's thinking shows how far he was from the theory that Charles Darwin had already begun to formulate in secret (Hodge, 1972). Life had emerged on the early earth by spontaneous generation, and like Lamarck, Chambers saw electricity as the "vital spark" that made this possible. Again like Lamarck, he then saw life advancing steadily up a linear chain of being toward the human form. This took immense periods of time, of course, but there was now enough of the fossil record available for it to be argued that the outlines of the process were substantiated by hard evidence—even if there were still many pieces missing from the sequence. Where Chambers differed from Lamarck was his almost complete lack of interest in adaptation. There was nothing in his theory comparable to Owen's (and Darwin's) vision of adaptive radiation from a generalized ancestral form.

The one thing that *Vestiges* did not seek to conceal was the most controversial implication of the theory—that humans had appeared as the last step in the progressive sequence. Chambers was quite explicit that the theory had to be applied to the mind as well as the body. The human soul could not be seen as a latecomer miraculously plugged in to a naturally evolved body. He appealed to the popular science of the mind known as phrenology, according to which each faculty of the mind was associated with a particular section of the brain (Cooter, 1985). The phrenologists were sneered at as pseudoscientists by the conservative elite because they claimed that they could "read" someone's personality from the structure of their skull. Yet their theory contained the germ of our modern efforts to chart how mental functions are localized in particular parts of the brain. For Chambers, the theory offered an explanation of how evolution could account for the superiority of human mental functions over even the higher animals. If transmutation was programmed to increase anatomical complexity, and if this applied to the brain, then the addition of extra cerebral structures would account for the higher mental faculties.

Vestiges sold well and became something of a popular sensation. According to James Secord (2000), it was by reading this book that the public was introduced to the idea of evolution. In effect, *Vestiges* made the Darwinian revolution possible, because it ensured that the *Origin of Species* did not come as a bolt from the blue. More seriously, the book also shaped the way people would read the *Origin*. Darwin's was not a theory of inevitable progress, but many assumed that it was because their thinking had been preconditioned by Chambers's book. If Secord is right, then at the popular level, Darwin merely finished off the revolution in thought that Chambers had already begun.

The one area in which *Vestiges* did not succeed was in changing the attitude of the elite scientific community. Those of a conservative disposition hated the book for what they perceived to be its materialism—all that talk of divinely planned laws was a smokescreen as far as they were concerned (Gillispie, 1951; Millhauser, 1959). Geologists such as Adam Sedgwick and Hugh Miller focused on the fossil record, where there were still many gaps that could be interpreted as evidence of sudden creations. But it was the book's implications for the human soul that were the real source of annoyance. Miller, a staunch Scots Presbyterian who had raised himself from working-class origins thanks to his discoveries of fossil fish, was quite explicit about this. In his *Footprints of the Creator* (reprinted 1850) he admitted that if design were the only problem, he might be able to accept that new species appeared by divinely implanted laws rather than by miracle. But if that meant that the human soul were merely the extension of animal mentality, then transmutation was unacceptable in this one instance, and by implication right across the board.

The refusal of conservative religious thinkers to take *Vestiges* seriously is hardly surprising. What was more important in the long run was that those with a more liberal interpretation of their faith were increasingly willing to accept the argument that Chambers had—perhaps rather cynically—put forward. Maybe God's designing hand was best seen in the operation of laws rather than in arbitrary miracles, and if so, why not admit the possibility that He had programmed a law of development into nature that would unfold toward the living things of today? This position was openly endorsed by the Oxford mathematician Baden Powell (Corsi, 1988b). Perhaps an exception might have to be made in the case of the human soul—both Lyell and Owen, for instance, shared Miller's reservations on this point (see Bartholomew, 1973 on Lyell's difficulties with evolutionism). But both were also increasingly willing to contemplate the idea of evolution as the unfolding of a divine plan. Few scientists were as yet willing to openly support transmutation, but there seems little doubt that in the course of the 1850s attitudes were changing in a way that made it easier for Darwin to come out into the open.

If the exponents of design were less inclined to endorse the old idea of miraculous creation, Darwin was also aware that a new generation of scientists was emerging that had less patience with natural theology, let alone with miracles. The best example of this new generation is the man who would later be known as "Darwin's bulldog," Thomas Henry Huxley (Desmond, 1994). Huxley came from a poor background and had used a medical training to gain experience as a comparative anatomist. Like Darwin he traveled the world aboard a Royal Navy survey ship, but where Darwin was the captain's companion on the *Beagle,* Huxley was a lowly naval surgeon on HMS *Rattlesnake.* Still, he had made a reputation for himself describing and classifying the exotic marine creatures they had dredged up, and in the 1850s he was desperately struggling to establish a career as a professional scientist. This was still not easy—there were few properly paid jobs, and Huxley was lucky to get a lectureship in paleontology at the newly established Royal School of Mines. Once secure, Huxley threw himself into the campaign to establish science as the main source of expertise that the government of an industrial country should call upon to solve its social problems. The traditions that had subordinated the life sciences to the Church via natural theology represented everything that Huxley wanted to change, so it is not surprising that he was on the lookout for radical theories that would undermine the design argument. Huxley came from exactly the milieu that found Chambers's popular science of progress appealing. But Huxley had already progressed beyond Chambers in his thinking. He wrote a viciously critical review of a late edition of *Vestiges* in 1853. Vague laws of progress shaped by a divine plan were just the kind of compromise with natural theology that he wanted to discredit. When Huxley jumped to evolutionism, he would want it to be a completely natural process—which is why he would be attracted to Darwin's theory as soon as it was presented to him.

Why did Huxley not appeal to the one theory of natural transmutation that was already in circulation, Lamarck's inheritance of acquired characters? This had been heavily discredited by Owen—Huxley's arch-rival—and he would not have wanted to associate himself with a theory known to have been favored by radicals. But there was another figure who had no such misgivings. This was the philosopher and political writer Herbert Spencer. Like Huxley he came from a poor background and worked for a time as an engineer before gaining a reputation through his writing. Spencer was the apostle of free-enterprise individualism, which he saw as the driving force of social and economic progress. Like Chambers, he saw how a theory of biological progress could underpin a campaign for social progress in a rapidly industrializing economy. But unlike Chambers and Huxley he realized that the inheritance of acquired characteristics offered just the mechanism for the job. If individuals struggled to improve themselves, they would improve their society too—and if Lamarck was correct, they would pass those improvements on to the next generation. Progress wasn't the result of a mysterious law operating behind the scenes, it was the product of millions of acts of individual self-improvement over many generations. Already in 1851 Spencer had written openly in support of Lamarck's theory, and in 1855 his *Principles of Psychology* explained the faculties of the human mind as the accumulated effect of learned habits transformed into inherited instincts (the inheritance of acquired *mental* characteristics). Had Darwin not published, Spencer would certainly have made his own bid to force the scientific community to take a new look at the question of transmutation.

By the late 1850s the world had changed in a way that would make it possible for the *Origin of Species* to make an immediate impact. No educated person still thought that the Genesis story offered a literal account of creation—at the very least there must have been a succession of divine creations. Thanks to *Vestiges*, everyone knew that evolutionism offered an alternative that might not require its supporters to identify themselves as atheists and materialists. Even some conservative scientists like Owen were conceding that the divine plan might unfold by law rather than miracle. Admittedly, the liberal theology underlying these moves was taking these thinkers a long way from the traditional Christian view of human sinfulness and the need for redemption. Chambers's Designer was at best the God of the deists, a God who did not interfere with the universe He had created. The more the scientists replaced miracle by law, the more the argument from design became associated with a liberal theology that saw the striving for progress in this world as a human duty.

The potential dangers of this move, as far as conservative Christians were concerned, were all too apparent when one looked at the very similar position being developed by the radicals, who saw no point in retaining a role for religion in modern life. Huxley and the new generation of openly naturalistic thinkers were looking for a new initiative in science that would allow them to discredit the argument from design completely. As yet they distrusted non-scientists like Spencer, who were trying to revive Lamarck's ideas (although, as we shall see, Lamarckism would still have an important role to play). But Spencer was growing in influence outside science as the rising middle class looked for an ideology that would make their own efforts appear to be the key to social progress. Progress was the key—but was the course of progress governed by a divine plan, or was it purely the result of human initiative? Whichever position one took on this divisive issue, there was an expectation that the old idea of miraculous creation would have to be rejected. The human soul would become a product of natural evolution, not a divinely implanted spiritual element transcending the material world. All the ingredients of an explosive mixture were present. What was needed was a spark that would set the whole thing off—and the spark would have to overcome the one remaining barrier to acceptance of evolutionism, the reluctance of the scientific community to embrace such a radical alternative to the old worldview. It would be Darwin who would supply this spark.

CHAPTER THREE

DARWIN AND HIS BULLDOG

Charles Robert Darwin conceived his theory of evolution by natural selection in the late 1830s and developed it in secret for the next two decades. When finally revealed in his *Origin of Species* of 1859, it reopened the issues already identified in Chambers's *Vestiges* and provoked a new and equally bitter debate. But there were significant differences this time. Darwin was an established scientist and his theory was widely acknowledged—even by some who rejected it—as a legitimate hypothesis. As a result, the debate panned out very differently. The elite of the scientific community had stood firm against *Vestiges*, but now they slowly swung round to concede that evolutionism, in some form at least, was acceptable. Conservative religious thinkers continued to oppose the idea, but they were increasingly challenged. In the popular mythology created by Darwin's supporters, the confrontation in 1860 between Thomas Henry Huxley and Samuel Wilberforce, the bishop of Oxford, symbolized the triumph of the Darwinians over their religious opponents. Soon even liberal theologians were jumping onto the bandwagon of evolutionism, while secularists applauded Huxley's efforts to replace the argument from design with an explanation based on natural law.

The situation was really more complex than it has been made to seem in hindsight. In some respects this first debate over Darwinism defined the issues that still trouble fundamentalists today. Darwin's theory was very different to Chambers's—indeed the latter's vague "law of development" was hardly a scientific theory at all. With natural selection there is no room for even an indirect form of design. The laws of nature operate without concern for future goals, and with an apparent disregard for the well-being of individual organisms. New characters appear more or less at random and are whittled down by a merciless struggle for existence to leave only those with survival value. This is evolution by trial and error, not by design. It produces species adapted to the environment, but there is no drive to perfection and no trend leading to humans as the goal of creation. As conservatives such as Wilberforce realized, this was a theory almost impossible to reconcile with traditional religious values, especially if the human mind is seen as a product of such an apparently undirected process.

Darwin thus anticipated almost all of the factors that allow modern atheistic Darwinians such as Richard Dawkins to exploit the theory in their campaign against religion. Dawkins seems to continue a tradition founded by Huxley almost as soon as the *Origin of Species* was published. Modern creationists are merely defending the position adopted by Wilberforce. In fact, though, there are major differences between the earlier clashes and those that reignited in the mid-twentieth century. For a start, Huxley was not entirely enthusiastic about the theory of natural selection. He recognized it as a valid scientific hypothesis, but he didn't think it was adequate to explain how evolution worked. Nor indeed did the vast majority of those who called themselves Darwinians. As we shall see in the next chapter, scientists proposed several alternatives to natural selection, and it was only with the advent of genetics after 1900 that the selection theory began to emerge as the most plausible explanation of evolution. The early Darwinians used that label for their position because they had followed Darwin in accepting the general idea of evolution, not because they were enthusiastic converts to the selection theory. In this respect the definition of Darwinism has changed significantly, because now it denotes Dawkins's position that natural selection is the only mechanism of change.

Because most of the early Darwinians were not rigid selectionists, they didn't confront all of the issues that disturbed conservative religious thinkers. In fact, as the historian James R. Moore (1979, 1985a) notes, there seems little difference between secular Darwinists and liberal theologians, who also accepted evolutionism. Both shared the assumption that evolution was progressive and purposeful, driving life up the scale of mental and physical development until eventually the human race appeared. Humans were the goal of progress—which didn't seem all that different from Chambers's claim that we are the goal of the divine plan of development. To make this assumption of progress work, however, the first generation of Darwinists had to evade the radical implications of the selection theory, which the conservatives had—quite rightly— identified as difficult to reconcile with religion. Natural selection was accepted as a negative process that weeded out the less successful of nature's products, but could not generate the new characters that led to progress. There was a feeling that nature must produce new characters in a more directed manner than Darwin's "random variation" implied.

In this respect, then, the earlier debate was not a rehearsal for the modern conflicts. Darwinism was seldom promoted in its most radical form, and the apparently black-and-white alternatives that confront us today were blurred by a spectrum of intermediate positions, all of which were designed to make evolutionism seem compatible either with a liberal theology or with a secular ideology equally dedicated to the idea of progress. Only when twentieth-century biologists demolished the plausibility of the alternative mechanisms of evolution did the more radical alternative of natural selection make itself felt. At the same time, the faith in progress common in the nineteenth century was undermined by catastrophes such as the Great War. To understand the original debates over Darwinism we must account for the very different cultural environment into which the *Origin of Species* was launched.

This chapter will first look at Darwin's career, identifying how he made the general case for evolutionism and the more specific case for natural selection as the mechanism of change. This will reveal which of those aspects of the selection mechanism made it a threat to the argument from design. It will also show why this approach to evolution made it difficult to see the human race as the goal of creation. We shall look briefly at Darwin's gradual loss of faith in Christianity, asking whether that was driven by personal factors or by recognition of his theory's implications. We shall also look at the claim, widely advanced by left-wing thinkers, that natural selection was modeled on the ideology of free-enterprise individualism. We might expect this effort to undermine the scientific credibility of the theory to appeal to modern creationists. In fact, though, they have been reluctant to endorse an argument that originates from a political viewpoint they reject.

Later in the chapter we look at the debate sparked by the *Origin of Species,* focusing on the cultural and social factors that determined how people reacted to the theory. We shall see how the theory was used as a weapon in an existing ideological battle between Huxley's generation of newly emerging professional scientists and the old guard, who sought to keep natural history subordinate to religion. We must also distinguish between conservative theologies which sought to highlight and oppose the more radical implications of Darwin's theory and liberal positions which softened the blow by implying that evolution must be progressive. By accepting that God's plan allowed the human race to perfect itself in this world, however, the liberals abandoned the traditional faith in which the sacrifice of Jesus Christ was the only thing that could save a sinful humanity.

THE DEVELOPMENT OF DARWIN'S THEORY

Charles Robert Darwin was born in 1809, the son of a prosperous medical doctor who had also made successful investments in the new industrial economy (for biographies see Browne, 1995, 2002; Bowler, 1990; Desmond and Moore, 1991). His grandfather was Erasmus Darwin, the author of the evolution theory outlined in his *Zoonomia*. The family had close links with the Wedgwoods, whose pottery firm was one of the great success stories of the industrial revolution. Darwin would eventually marry his cousin, Emma Wedgwood. The family background thus identified Darwin closely with the rising class of entrepreneurs who were creating the new industrial capitalism.

The young Darwin acquired an early interest in natural history, which he retained when he was sent as a medical student to Edinburgh. Here he met the Lamarckian anatomist Robert Grant, and although he later claimed to have been unaffected by Grant's radical ideas, the historian Philip Sloan (1985, 1986) has shown that he did take an interest in issues defined by an evolutionary program. He was fascinated by Grant's work on zoophytes (corals, etc.), which were seen as a bridge between the animal and plant kingdoms. But Darwin's medical career was short-lived. He hated the practical side of medical study and left Edinburgh for the far more conservative atmosphere of Cambridge. An Arts degree at Cambridge was often the prelude to taking holy orders in the Church of England. At this time Darwin's religious views seem to have been quite orthodox, and he was happy to envisage a career as a country vicar studying natural history on the side.

Darwin did not study science at Cambridge, but he formed extracurricular links with the professor of geology, Adam Sedgwick, and the professor of botany, John Stevens Henslow. Working with them gave him a good training in natural science, and before he graduated Darwin decided that he wanted to work full time on natural history in the hope of being accepted into the elite scientific community. Fortunately, the family's wealth meant that he did not have to seek a paid position as a scientist (because there were very few available, as Huxley would later discover). Darwin accompanied Sedgwick on geological expeditions to Wales, and we must remember that his early career was focused as much on geology as on natural history (Herbert, 2005). At the end of his time at Cambridge he was offered the opportunity that would change his life: the chance to travel on the survey vessel HMS *Beagle* (Moorehead, 1969).

The *Beagle* was being sent out for the second time to complete the British Admiralty's project to chart the coast of South America. Darwin was thus participating indirectly in the expansion of British imperial power—the country had no colonies in South America but traded extensively in that part of the world, and her ships needed accurate maps. It was normal for such surveys to include an element of natural history—usually the ship's doctor would collect and preserve specimens of animals, plants, fossils, and minerals from the countries visited. But the *Beagle*'s captain, Robert Fitzroy, wanted a gentleman companion on board—he had brought the ship home from her earlier voyage after the original captain had gone mad from stress and isolation. He made it known that there was an unpaid position on the ship for someone with the right training in science and the right social background. After overcoming doubts expressed by his father, Darwin was accepted. He would be away from England from 1831 to 1836.

The voyage of the *Beagle* transformed Darwin's view of the world and made his reputation as a scientist. He was not confined to the ship and made several expeditions to the interior of South America. His discoveries there converted him to Lyell's uniformitarian geology and provided him with the basis for scientific publications in that field. More important in the long run, a whole series of discoveries in natural history led him to doubt the plausibility of the creationist viewpoint he had absorbed in the conservative atmosphere of Cambridge. On the Galapagos islands, he would see examples of geographical variation which provided him with the evidence for a model of evolution based on the divergence of isolated sub-populations derived from a common ancestor (for a collection of important studies of Darwin's science see Kohn, 1985).

Sedgwick had trained Darwin in the catastrophist tradition, which explained the transitions between the geological periods in terms of violent upheavals. This was the approach that Lyell challenged in his *Principles of Geology.* Henslow gave Darwin the first volume of the *Principles* to take with him on the voyage, but warned him not to believe it. But Darwin's discoveries in South America soon converted him to the uniformitarian position. He observed a major earthquake in Chile, and noted how the whole land surface had been elevated, establishing a new shoreline along the coast. Stretching up the sides of the Andes was a series of ancient beaches, showing that the mountain range had been built by a series of earthquakes similar to those still occurring. There was no need to invoke a violent catastrophe to explain the uplift of the mountains. Later on Darwin developed a theory to explain the formation of coral reefs around Pacific islands on the assumption that here there was an equally gradual subsidence of the earth's surface.

Converting to Lyell's uniformitarian geology put Darwin in a position from which he could envisage changes in the organic world taking place equally slowly and gradually. He discovered fossils in South America that showed a relationship to the present inhabitants—giant sloths and armadillos. Evidently there was continuity in the inhabitants of the continent over geological time. Advocates of design could only explain this by invoking a "center of creation" specializing in these forms, but there was no obvious reason why the Creator should operate in this way. Darwin eventually came to realize that the phenomenon could be explained on the assumption that geographical isolation preserved the basic character of the original inhabitants while they underwent superficial modification adapting them to changing conditions.

On the pampas, the open plains of Patagonia, Darwin discovered that there were two species of the ostrich-like bird, the rhea. Each had its own main territory, to which it was well adapted, but the two populations overlapped in an intermediate region. Neither was fully adapted here, making it difficult to apply Paley's notion that each species was perfectly adapted to its home environment. Observing the conflict between white settlers and the native Indians, Darwin could imagine that the two species of rhea were competing to occupy as much territory as possible. Here was the model for a much less harmonious view of how species were related to their environment. If the conditions changed slightly as the land was modified by geological forces, one might expand its territory and eventually drive the other to extinction. From the start, Darwin's vision of the history of life included as much room for extinction as for evolution. When he came to see evolution as a process in which new forms branched out from an original ancestor, he knew that there was always room for new species because extinction was constantly pruning the existing branches of the tree of life.

It was on the Galapagos islands, off the Pacific coast of South America, that Darwin made the observations which converted him to a belief in evolution. It used to be assumed that he underwent a kind of "eureka" experience when he saw the implications of the geographic variation among the birds now colloquially known as "Darwin's finches." There were many different forms of finch on the various islands, each adapted to a particular way of feeding. Darwin would eventually realize that each of these forms could have evolved from a single ancestral type, specimens of which would have been accidentally blown across from the mainland by storms. On each island a small population would have been established and, developing largely in isolation, had adapted to its new conditions in its own way. Thus from a single original form, geographical isolation and adaptation had produced a group of closely related descendants.

We now know that there was no "eureka" experience, and that Darwin very nearly missed the significance of the finches altogether (Sulloway, 1982). It was only just before the *Beagle* left that he was informed that the giant tortoises after which the Galapagos are named showed distinctive characters on each island. He then began to re-examine his specimens of other species and belatedly realized the extent of the geographical variation. The mockingbirds were probably more important at first than the finches, even though the latter have become the iconic example of the phenomenon. Only when he returned to England did Darwin learn from expert ornithologists that the birds he had collected would have to be regarded not as local varieties of a single species, but as distinct species in their own right.

He was now faced with a dilemma: to preserve the traditional view that each true species was separately created, he would have to accept that God had performed a separate miracle for every finch and mockingbird species on each of these insignificant islands. He decided that this position reduced the creation hypothesis to absurdity. He might have adopted the view, favored by most modern creationists, that God only creates the basic types and has permitted a certain amount of low-level speciation as the original population spreads out. But Darwin was not prepared to accept such a compromise. Shortly after his return to England, and while making a name for himself by publishing his geological findings, he became a full-scale transmutationist. Behind the scenes, he began to explore the implications of the possibility that all species have been produced by natural causes from previously existing ones, the tree of life extending so far back into the past that even the major branches would at last merge together.

Darwin's notebooks (now published as Darwin, 1987) show how he developed this idea in the late 1830s. He was convinced that the major problem was to explain how species are modified to become adapted to changes in their environment. In this sense, Paley's emphasis on adaptation (which had impressed Darwin as a student at Cambridge) still shaped his thinking. Could there be a natural process that would produce the structures we have hitherto attributed to divine benevolence? Lamarck had posed this very question and had answered it by proposing the mechanism of the inheritance of acquired characters, described in Chapter 2. Darwin was not convinced that this was an adequate explanation, although he always accepted that Lamarckism might play a subsidiary role. Unlike Lamarck he accepted the reality of extinction, and he saw evolution as the branching off of many differently adapted forms from a common ancestor, not the ascent of a ladder of progress.

Darwin sought an alternative explanation of how adaptive modifications could be produced. Following Lyell's dictum that "the present is the key to the past" he looked for clues in the variability of modern species. He began to study the work of animal breeders and horticulturalists, who were known to be able to produce significant new characters in the species they worked with (on the development of the selection theory see Hodge, 1985; Hodge and Kohn, 1985; Kohn, 1980; Sloan, 1985).

There were plenty of enthusiasts Darwin could talk to. Some bred fatter cattle and sheep to feed the growing population. But others bred for mere decoration, as with fancy varieties of dogs and pigeons. From them Darwin learned that any population consists of a collection of differing individuals. Just like human beings, every individual has his or her own peculiar character. We may think that all members of a wild population are identical, but anyone who looks closely will see small individual differences. Darwin began to see how the breeders exploited these small variations to produce massive differences in the animals they worked with. The individual differences seemed to appear at random, not in the sense that they were uncaused, but because whatever produced them generated a variety of differences with no regard to what was good for the species. Today we see the differences as a consequence of the genetic diversity of the population. Darwin didn't think in terms of genes, and instead speculated that a changed environment might somehow upset the reproductive process to produce minute differences between parent and offspring.

The key to the breeders' success was a process of selection. In nature, the different individuals breed promiscuously and their characters are simply mixed up. But the breeder who is attempting to produce something new—say a dog with longer legs to run faster—looks at the young in each litter and picks out those that have a slight variation in the right direction, in this case those with the longest legs. Only these are allowed to breed, so the next generation is composed solely of the offspring of the longest-legged dogs, and they will all have legs of above-average length. The selection process is then repeated over many generations, and the results can be quite spectacular, as witnessed by the many breeds of dogs. Some of these show characters so different that in nature they might have been classed as distinct species. The one difference is that in varieties produced by artificial selection, interbreeding is still possible, so the different forms still belong to the same species.

Darwin now began to wonder if there could be a natural equivalent to this process of artificial selection. The breeder interferes with the natural process of reproduction, but could there be a natural form of selection that could similarly bias reproduction in favor of those with a particular character? Specifically, could there be something that would promote the breeding of those individuals that best fit the environment while suppressing the reproduction of those less well adapted? Darwin realized that such a pressure could be exerted by what he called the "struggle for existence," borrowing the phrase from the writings of the economist Thomas Robert Malthus. Darwin read Malthus's *Essay on the Principle of Population* at a crucial point in his research, and from it deduced

that in every species there was a tendency for the population to expand. (I discuss Malthus in more detail later in this chapter.) Since
resources are constant, there must be a competition to determine
who gets enough to survive and reproduce. Darwin's great insight
was that any individual born with a slightly favorable variation
(making it better adapted to the environment) will have a better
chance of succeeding in the struggle for existence, and hence a
better chance of reproducing. Those with harmful characters will
be less likely to survive and reproduce. Thus there will be a natural
process of selection tending to enhance any character that adapts
the population to changes in the environment. Here is Darwin's de-
scription of how the process would work to produce a long-legged
form of dog, from the essay he wrote describing his theory in 1844:

> . . . let the organization of a canine animal become slightly
> plastic, which animal preyed chiefly on rabbits, but some-
> times on hares; let these same changes cause the number of
> rabbits very slowly to decrease and the number of hares to
> increase; the effect of this would be that the fox or dog would
> be driven to try to catch more hares, and his numbers would
> tend to decrease; his organization, however, being slightly
> plastic, those individuals with the lightest forms, longest
> limbs and best eyesight (though perhaps with less cunning or
> scent) would be slightly favoured, let the difference be ever so
> small, and would tend to live longer and to survive during
> that time of the year when food was shortest; they would also
> rear more young, which young would tend to inherit these
> slight peculiarities. The less fleet ones would be rigidly de-
> stroyed. I can see no more reason to doubt but that these
> causes in a thousand generations would produce a marked ef-
> fect, and adapt the form of the fox to catching hares instead of
> rabbits, than that greyhounds can be improved by selection
> and careful breeding (Darwin and Wallace, 1958, p. 120).

This was the idea that Darwin had put together by 1840 and developed largely in secret for the next twenty years. The 1844 essay was not intended for publication, but was written because Darwin was now suffering from a chronic illness and feared that he might die. It is easy to see why he was so cautious: natural selection would replace design with random variation and a merciless struggle for existence that eliminates all but the lucky few. Given the outcry over Chambers's *Vestiges of Creation,* which proposed a divinely implanted law of progress, we can see why Darwin feared that his own suggestion would be greeted with even more cries of outrage. If natural selection were the only mechanism of evolution, there was no guiding hand behind the process, no benevolent Designer, and no trend forcing life automatically toward higher levels of organization or toward humanity. Darwin was certainly aware of these consequences, but he also held back from publication because he needed more information to substantiate his theory.

Darwin may not at first have realized how subversive his idea was. In the 1840s he still hoped that his theory might be compatible with the belief that the laws governing the process were instituted by a wise and benevolent God (Ospovat, 1981). By the 1850s his views had changed, however, and his vision of nature became darker, his views of religion less positive. In part the change was personal—the death of his beloved daughter Annie in 1851 left him with questions about the existence of a caring God (Moore, 1989a). But there was also a growing realization that if Malthus was right, then the struggle for existence went on relentlessly, however well the species might be adapted to its environment. Nature was a scene of constant death and endless selfish struggle, hardly the sort of process one would expect a benevolent God to establish.

The central role played by the idea of population pressure also highlights Darwin's reliance on an idea derived from a political ideology. He was convinced that humans would have to be included in the evolutionary system, and had begun to think about how our intelligence and our social instincts might have been shaped by natural selection. He was now a materialist, convinced that the mind was dependent on the activities of the brain, and that moral values were merely rationalizations of the instincts natural selection imposed on any species that lived in social groups. There was no reason why he should not be prepared to extend an idea developed in the context of human society to the natural world. But Malthus's principle of population had an ideological foundation. It was presented as a contribution to what was known as "political economy," an attempt to uncover the laws governing both society and the economy. In the context of early nineteenth-century Britain, this meant understanding the working of the new system of free-enterprise capitalism. Malthus challenged the reformers who thought that poverty and starvation could be eliminated by government action. He argued that poverty was not the result of an artificial social hierarchy—on the contrary it was both natural and inevitable. Given the "passion between the sexes" there would always be too many children born for the food supply to support, with the result that some must inevitably go hungry. State support for the poor should be abandoned, because it was better to let a few starve now rather than allow the population to expand unchecked to a level where mass starvation was inevitable.

Malthus thus endorsed free-enterprise individualism, the system of *laissez-faire* (the notion that the state should not interfere with the activities of individual citizens). The fact that Darwin drew upon Malthus to justify his claim that there would always be a struggle for existence lends support to the criticism that his theory merely projected capitalist values onto nature. Karl Marx commented on the close parallels between natural selection and the capitalist vision of competition as the spur to economic progress. The political Left routinely argues that Darwin's theory is bad science—far from being based on a study of nature, it creates an artificial model of the natural world based on a particular value system.

Historians such as Robert M. Young (1985) use the link with Malthus to support this claim, and the biography of Darwin by Adrian Desmond and James R. Moore (1991) also sees his thinking as deeply influenced by the social debates of the time.

Malthus did not advocate struggle as a means of distributing wealth or as the driving force of progress (Bowler, 1976b). He thought wealthy people gained their riches by inheritance, not by their own efforts, and used the term "struggle for existence" only when describing the warlike tribes of Central Asia. Nor did he think that free enterprise would generate progress—his whole purpose was to argue against the idea of progress. Whatever Darwin's debt to Malthus, he went much further in seeing the struggle for existence as a universal process capable of changing the nature of a species.

Nevertheless, the claim that the selection theory is modeled on a political system has implications for its scientific validity. Surely science should be based on the study of facts, not on human values. But such an argument has less impact when we accept that science always advances through the proposal of hypotheses which are then tested against the facts. Scientists have been inspired by a variety of sources when thinking up their models of nature, including art, philosophy, politics, and religion. If the hypothesis proves fruitful in generating research, then it is doing its job in science. The socialists claim that Darwin's theory doesn't provide such an impetus because the testing process itself is warped by the preconceptions of those who do the testing. It seems curious that modern creationists ignore this argument against Darwinism's scientific credentials, given that they are always on the lookout for anything that will discredit the theory. But when we remember the source of this attack, we can see why it seems unattractive to those from the Religious Right who are convinced that the free-enterprise system is a God-given model for how we should govern ourselves. The claim that the theory they despise is a reflection of their own political values is more an embarrassment to them than an opportunity.

From the late 1830s Darwin set out on an extended project to explore his theory's implications and to test them against the best available evidence. He began to correspond with a wide range of naturalists, seeking information that he could use in his investigations. A few were eventually informed of the true nature of his project, including the geologist Lyell and the botanists Joseph Hooker and Asa Gray. It was only at a later stage that he contacted the man who would become identified as his closest disciple, the zoologist Thomas Henry Huxley. All of these contacts helped Darwin to show how biology would be transformed by the idea of evolution. Hooker debated with Darwin on the question of the geographical distribution of plants before he was converted. Darwin himself undertook a survey of the living and fossil barnacles, which started from some strange specimens brought back from the *Beagle* voyage. This gained him new insights into the problems of adaptation, classification, and the relationships between embryonic and adult forms. No one would be able to accuse Darwin of being an inexperienced naturalist who did not understand the extent of the problems to be tackled.

By the 1850s Darwin felt that he had enough evidence to put a convincing case before the scientific community. He also sensed that the atmosphere in the country was changing. As people became more accustomed to thinking about the issues raised by *Vestiges,* they gradually became less frightened at the prospect of a history of life based on natural law rather than divine miracle. Perhaps God's plan of creation did unfold gradually, rather than through a series of discrete supernatural interventions. Darwin began to write his ideas up into what would have been a massive two-volume survey (parts subsequently published as Darwin, 1975). But he would never finish it. The project was interrupted by the arrival in 1858 of a paper on natural selection written by Alfred Russel Wallace.

Much has been written on the apparent coincidence of two British naturalists independently discovering the same theory. In fact,

Wallace's ideas were significantly different from Darwin's, in part because they came from different social backgrounds and had different scientific interests (Kottler, 1985; Fichman, 2004). Wallace was struggling to support himself by collecting rare animals in the tropics. He was deeply religious and a lifelong campaigner for social reform. He did not appreciate the analogy between natural and artificial selection that so fascinated Darwin, and he initially conceived of selection acting between rival subspecies rather than between individuals in the same population. But his 1858 paper—written on an island in what is now Indonesia and sent to Darwin for comment—showed clear parallels with the theory that Darwin had been working on for twenty years. Darwin panicked and called in Lyell and Hooker for advice. They suggested that he arrange for the publication of the paper, along with a brief account of his own theory demonstrating priority (reprinted in Darwin and Wallace, 1958). He then rushed to complete the more detailed account we know as the *Origin of Species.*

THE ARGUMENT OF THE *ORIGIN*

When Darwin's book was published in late November 1859, 1,250 copies were sold to the bookshops on the first day. It was the *Origin* that sparked the great debate and paved the way for the conversion of the scientific community—and much of the educated public—to evolutionism. We should examine its argument in some detail so we can see which aspects of Darwin's case were found most compelling.

Darwin's book had three interlinked purposes. Most immediately, it developed the case against a simple creationism linked to Paley's version of design. Most of these arguments are ignored by modern creationists. More constructively, it argued that many otherwise puzzling aspects of natural history can be explained if we postulate a form of evolution in which populations derived from a single original species can become separated and will then evolve in different directions, driven by the need to adapt to their different environments. This is the basic theory of common descent, or divergent evolution, and it marks a major improvement on the old idea of a linear scale of development aimed at a single goal (the human race). Finally, at the most detailed level, Darwin wanted to convince his readers that his theory of natural selection offered the best explanation of how this adaptive evolution occurred.

Surprisingly for the modern reader, the book begins with the argument for natural selection, leaving the general case for evolution until later. This was because Darwin saw natural selection as his crucial initiative—a new, fully scientific explanation of evolution which would lift the general theory from a vague speculation to something that could be used as a basis for research. This was the breakthrough that Darwin hoped would force naturalists to reconsider the whole situation. But for the sake of convenience, let us invert Darwin's order of business and begin with the case against creation or design and in favor of divergent evolution.

Darwin had originally supported Paley's version of the design argument, and natural selection was intended to replace divine contrivance as an explanation of adaptation. There were problems both with the idea that species are perfectly adapted to their environment and with the claim that each is a divinely created unit. Darwin noted the imperfection of many adaptations. Our own poor adjustment to upright walking is the source of the back pains that plague many people, and the inappropriate relationship between the esophagus and the windpipe leaves us vulnerable to choking on our food. There are species of birds with webbed feet that never go near the water. The idea of perfect adaptation doesn't work at the detailed level required by the naturalist. Species are loosely adapted to their way of life and to their environment, but the imperfection of the fit suggests that there is a *process* going on: species move into new environments and then begin to adapt to them. In the case of more fundamental problems like the human back, it looks as though nature cobbled together adaptations from fundamentally unsuitable starting material. If there is a designing God, then He has chosen to limit His activities by always modifying a pre-existing form even where that wouldn't have been the best starting point. All of this makes more sense if adaptation is a product of natural evolution, rather than of divine contrivance.

On the fixity of species, Darwin could appeal to the many problems that had been encountered by naturalists trying to apply this idea in practice. At first sight it looks as though nature is divided into discrete units that we might take to be species, each descended from an originally created pair. Traditionally, it was asserted that even closely related species cannot interbreed. But at the detailed level this simplicity disappears. Naturalists constantly disagreed about whether some forms were separate but closely related species or merely local varieties of a single species. This shouldn't happen if God had created a collection of clearly distinct species. Even the sterility criterion breaks down in practice. Everyone knows that the horse and the ass can interbreed to produce a mule, and we have to appeal to the fact that the mule is normally sterile to maintain the claim that they are distinct species. But not all hybrids are totally sterile, even in the animal kingdom, and among plants successful inter-species hybridization is commonplace.

The fact that the distinction between species doesn't work in practice suggests that they are not unambiguous units in a divine plan—again there seems to be a process at work in which a single species can divide into a number of descendant forms that move gradually further apart until eventually they become distinct species. Many modern creationists accept that closely related species such as the Galapagos finches have been produced by divergence from a single, originally created type. The problem with this position is: who decides which types are really fundamental? Aware of the arbitrary nature of such a decision, Darwin argued that it would be better to abandon the whole idea of originally created starting points.

We are already moving toward the argument in favor of common descent. Evolution is not the ascent of a ladder, step by step, from the amoeba to the human form. The only diagram in the *Origin* is an idealized model of evolution as an ever-branching tree. Each branch continues for some time but eventually either comes to a dead end (it becomes extinct) or diverges into a group of related forms. The divergence is partly due to the need to adapt to new conditions when the original population becomes geographically fragmented, as with the Galapagos finches. But Darwin also realized that there was a tendency for species to become ever more specialized for a new way of life, even if their environment remained constant. The branching nature of evolution was one of Darwin's key insights, although his followers were constantly tempted to reimpose a ladder of progress. They were less sure that the only cause of change was adaptation, and far less convinced by his case for natural selection.

Once the branching nature of evolution is recognized, many otherwise puzzling aspects of the natural world start to make sense. The Galapagos finches are closely related because they have recently diverged from a common ancestor and share the basic characters of that ancestral form, with superficial adaptive modifications. But if that degree of similarity is a sign of common descent, why not apply the same argument to all the other degrees of relationship used to classify animals and plants? Common descent, coupled with frequent extinction, explains why the natural world can be divided into groups. The groups are signs of common ancestry, the gaps between them of the fact that many branches in the tree of life have gone extinct, wiping out the intermediates that the creationists think ought to bridge every gap in the system. The only way the creationist can explain degrees of similarity is to abandon the idea of perfect design and accept (as Owen had argued in the 1840s) that God has chosen to work to a plan which generates relationships exactly similar to those that would be predicted by the theory of common descent.

The later chapters of the *Origin* explore many other factors that start to make sense in the light of evolution. Why are embryos sometimes a better sign of the relationships used to classify species than the adult forms? The barnacles on which Darwin worked had originally been classed as mollusks, but were revealed as crustaceans as soon as their larval forms (the free-swimming young) were known. Darwin had little interest in what is sometimes called the recapitulation theory, in which the human embryo is supposed to mount through the sequence of its evolutionary ancestors as it develops. But he did recognize that embryos preserved deep ancestral forms better than the adults, which are more subject to adaptive modification. Geographical distribution becomes clearer when we realize that newly emerging groups can only spread out to occupy territory that is open to them by migration. That is why the inhabitants of South America and Australia are so distinctive—these landmasses have been cut off from the rest of the world during much of their history and have not received the immigrants that have wiped out older species in most of Eurasia and North America.

This moves us on to a major debating point: the fossil record. Anyone who appreciates that evolution is a branching process can see how pointless it is to ask why we don't see intermediates between all the living forms of today. But opponents of evolution (from long before Darwin's time) have always tried to argue that we ought to be able to trace the sequence of developments in the fossil record. Yet the record shows little continuity—it is full of gaps, which the creationists interpret, then as now, as indications that new species appear supernaturally. Darwin was aware of this objection and borrowed an argument from Lyell to defend himself. Lyell had pointed out that fossils are only preserved in certain rather unusual circumstances, where sediment builds up gradually on the bed of a sea or lake. Given that these circumstances only occur sporadically in time and place, there is no reason to expect that the fossil record will be a continuous record of all the changes that take place in the history of life. There will inevitably be many gaps, some of which we may fill in with future discoveries, but some of which will be permanent because the missing species were never fossilized. What we have to do, Darwin argued, is look at the record as it exists to see if—allowing for the gaps—the overall pattern of development is what we would expect from his theory. He could appeal to work by Richard Owen and others which had already demonstrated that the history of many groups is that of multiple lines of specialization radiating out from more generalized ancestors. Future research may reveal some of the missing links in the sequences, and we can predict what those links should look like, but Darwin himself was never optimistic about our ability to reconstruct all the details of the history of life on earth.

These arguments came in the later chapters of the *Origin,* following Darwin's detailed explanation of how the changes were produced through natural selection. We have already seen how Darwin pieced this theory together, but we need to think carefully about how the mechanism fitted into the overall argument. Natural selection implied an ever-present competition between individuals and between rival species occupying the same territory. It explained the drive toward increasing specialization revealed by the fossil record, because a more specialized form will always be more efficient at exploiting a particular way of life. But it also explained the constant threat of extinction due to environmental change or competition from rivals. Darwin built extinction into the very foundations of his thinking, thus undermining the optimism of Paley's vision of divine benevolence. The selfishness of natural selection also explains the existence of parasites—which are very hard to reconcile with the idea of a benevolent Creator.

Darwin was well aware that he was presenting a bleak picture of nature which left little room for reconciliation with the traditional idea of design. Natural selection worked only by adapting species to their environment. There can thus be no built-in progressive trend, no automatic ascent toward the human form as the goal of creation. Adaptation is bought at the price of the constant suffering of the unfit who must be eliminated in every generation. By the 1860s Darwin had accepted that there was no sense in which evolution could be seen as a process superintended by a wise and benevolent Creator. He was never a complete atheist, though, and continued to hope that there might be some indirect way by which evolution could be reconciled with the belief that the world has an ultimate purpose. It is worth noting that the story of Darwin's deathbed return to Christianity, frequently repeated by creationists, is a fabrication (Moore, 1994).

Darwin was unwilling to give up the prevailing faith in progress that formed the basis of middle-class ideology. There was no direct progressive trend—that was where Lamarck and Chambers had gone wrong—but natural selection could be seen as the driving force of a less structured form of progress. Darwin was also acutely conscious of the need to minimize the danger of a negative public reaction to his theory, and this meant allowing his readers to end on a positive note.

In the conclusion to the *Origin*, he thus waxed poetical over the long-range tendency to progress, which he believed lay behind the harshness of natural selection. He talked of life being "breathed" into primitive organisms, with the potential for endless improvement. He later regretted "truckling" to public opinion by using biblical language at this point—he certainly did not believe that the origin of life was miraculous—but it was necessary to distance his theory of how life evolves from the controversial idea that it

could have originated by spontaneous generation. From this starting point, he argued, progress was not only possible but inevitable,
at least in the long run.

> Thus from the war of nature, from famine and death, the
> most exalted object which we are capable of conceiving,
> namely the production of the higher animals, directly follows.
> There is a grandeur in this view of life, with its several powers,
> having been breathed into a few forms or into one, and that,
> while this planet has gone cycling on according to the law of
> gravity, from so simple a beginning endless forms most beau-
> tiful and most wonderful have been, and are being, evolved.
> (Darwin, 1859: 490)

Note how Darwin compares the laws of evolution with the laws of
planetary motion. In both areas, he implies, the laws of nature operate at a constant level. The difference is that in evolution, the law-
governed effects combine to give the process of natural selection,
which—however harsh it might seem at the individual level—is the
motor of universal progress.

THE GREAT DEBATE

There is a story that when the *Origin of Species* was published, a
clergyman pointed Darwin out as the most dangerous man in England. Conservative religious forces appreciated that here was a challenge they could not ignore. Thanks to Chambers's *Vestiges,* the idea
of evolution was in the air, and Darwin might now precipitate a
transition within the scientific community that would make it the
new orthodoxy. More seriously, here was a theory that challenged
even the compromise position floated by Chambers in which evolution was the unfolding of a divine plan. For all Darwin's efforts to
present his theory as a contribution to the ideology of progress, it
wasn't immediately apparent that this was the purposeful vision of

change that liberal religious thinkers might be tempted to accept. The more radical implications of the selection mechanism were frequently ignored, and Darwinism did become the scientific foundation for the ideology of progress. But to conservative religious thinkers the dangers were obvious from the start, and they made a concerted effort to discredit the theory.

Darwin's supporters confronted the conservative opponents, most visibly in the famous clash between Thomas Henry Huxley and Samuel Wilberforce, bishop of Oxford, at the 1860 meeting of the British Association for the Advancement of Science. We shall come to the story of that meeting in a moment, but it is necessary first to stress the need to go beyond the image of simple confrontation if we are to understand the impact of Darwinism on nineteenth-century thought. Huxley welcomed Darwinism in his fight against religion—yet he was never convinced that natural selection offered a comprehensive explanation of how evolution worked. Wilberforce was primed for his attack on Darwin by the anatomist Richard Owen, Huxley's great opponent in science—yet Owen supported the idea of evolution as the unfolding of a divine plan. Many of those who favored the ideology of progress resented the way in which it was being hijacked by radicals who refused to allow any compromise with traditional religion. The fact that Darwinism could be presented as a mechanism of "progress through struggle" opened the way for an alliance of liberal religious thinkers and philosophical supporters of free-enterprise capitalism to define a middle ground in which evolution transformed but did not destroy the old Protestant virtues (for surveys of the Darwinian revolution see Bowler, 2003; Himmelfarb, 1959; Ruse, 1979, 1996).

We start with the two opposite ends of the spectrum: Huxley and Wilberforce. We have already encountered Huxley, a rising star in the scientific community who found Chambers's *Vestiges* too much of a compromise with the argument from design (Desmond, 1994, 1997; Lyons, 1999). By the 1860s he was starting to make his mark and was actively promoting the role of the professional scientist in education and government. This was the basis of his opposition to formal religion, which was the traditional source of expertise and social authority and thus stood as a barrier to the advancement of the scientific community. Huxley favored Darwinism not because he was a complete convert to selectionism, but because he saw in the theory an example of how science could escape from the shackles of natural theology. For Huxley, nature worked solely according to laws that only scientific observation can reveal—there was no visible evidence of a supernatural world beyond. He did not deny the existence of a Creator who might have set the whole system going, and indeed coined the term "agnosticism" to denote a state of active doubt rather than of positive disbelief (Lightman, 1987). But he did insist that nature gives us no clues or hints about any ultimate purpose, so in that sense religion is an illusion. By definition, then, the origin of species must be a natural, not a supernatural process. Darwin's theory was another blow struck by science against religion. As Huxley said in his review of the *Origin:* "Extinguished theologians lie about the cradle of every science as the strangled snakes beside that of Hercules. . ." (Huxley, 1893–94, II, 52). He was very conscious that he was continuing a longstanding war in which science was seeking to take over territory once occupied by religion (Fichman, 1984).

Huxley was a very different kind of scientist to Darwin. Not only was he a professional rather than a gentleman amateur, he was also a representative of the new biology, which advanced in the dissecting room and the laboratory rather than out in the field. He thus approached the *Origin* in a specific way. He wasn't very interested in adaptation, and saw natural selection as a plausible hypothesis that might explain some aspects of evolution, but could not be the main mechanism. For Huxley, evolution offered a new way of thinking about how organisms were related to one another and how the various types of organization could have arisen. He would contribute to the evidence for evolution by applying his skills to the study of fossils, reconstructing the outlines of the history of life on earth and evaluating the new fossils that were being discovered as potential missing links. In the end, Huxley helped divert attention away from natural selection and the study of evolutionary mechanisms, creating an evolutionary project quite different from the one that Darwin had envisaged (Bartholomew, 1975; Bowler, 1996).

In the short term, however, Huxley's political agenda meant that he had to play a role in defending Darwin from the conservatives. If they saw natural selection as a threat, he would defend it, even if he didn't think it was that important. He wrote a review of the *Origin* in the influential London *Times* and another in the intellectual *Westminster Review*. And it was as a leading spokesman for Darwin that he came to the Oxford meeting of the British Association in 1860—Darwin himself was, as always, too ill to attend public debates. It was known that Bishop Wilberforce was going to attack the theory, primed by Huxley's arch-enemy, the anatomist Richard Owen. Huxley almost left before the big debate, but was persuaded to stay on by Robert Chambers, whose *Vestiges* had started the ball rolling fifteen years earlier.

Wilberforce's father, William, had led the campaign to abolish slavery in the British Empire. Samuel moved away from his father's evangelical position and established himself as a leading cleric in the conservative wing of the Anglican Church, just the sort of figure Huxley saw as a barrier to progress in the modern world. He was known as "soapy Sam" because of his eloquence. Like other conservative clerics, he was already involved in another controversy related to a more scientific way of thinking. This was over the emergence of the so-called "higher criticism," in which the text of the Bible was treated not as the Word of God but as a collection of ancient texts to be analyzed like any other. Originating in Germany, the implications of this approach had been highlighted in D. F.

Strauss's *Life of Jesus,* translated into English by Mary Ann Evans, better known as the novelist George Eliot. In 1860 a collection of articles written within this new tradition had appeared under the title *Essays and Reviews.* For conservatives like Wilberforce, those who moved in this direction were betraying the faith, giving their support to an approach that left the sacred text without the authority to defend the story of the creation or the miracles associated with Jesus's life and work. In some respects, this was a greater threat than Darwinism.

Even so, Wilberforce knew that he had to block the rise of evolutionism lest it lend support to those who challenged the validity of the Genesis text as an accurate picture of humanity's creation. He didn't know much about science, but Owen would provide him with ammunition to use against natural selection, which they both detested as rank materialism. Wilberforce's attack would focus on the two most disturbing aspects of the new theory: the challenge to design represented by natural selection, and the implications of a theory that depicted human beings as little more than improved apes. As far as he was concerned, it was simply inconceivable that a process starting from the random variations among individuals could generate anything equivalent to the designing hand of intelligence. Paley was right: the complexity and purposefulness of adaptive structures was beyond the scope of any natural process. Darwin himself conceded that Wilberforce's attack, which appeared in print shortly afterward in the form of a review of the *Origin of Species,* was a clever exposure of the weakest points in his argument.

But it was on the subject of human origins that Wilberforce sought to expose the materialism of the Darwinian position. Darwin had avoided all but a bare mention of this issue in the *Origin* precisely because it was so sensitive. But *Vestiges* had made it clear to everyone that a comprehensive theory of evolution must imply an animal ancestry for humankind, and Huxley was already arguing with Owen about the closeness of the relationship between humans and apes. To draw out the absurdity of such an ancestry, Wilberforce jokingly asked Huxley if he claimed to be descended from an ape on his grandfather's or his grandmother's side. According to the popular legend, this was Huxley's opportunity to crush the bishop. Some reports claimed that his reply was to the effect that he had rather be descended from an ape than a bishop. More likely, he said that he would rather be descended from an ape than from a man who misused his talents to attack a theory he didn't understand. Whatever his actual words, the result was uproar in the audience—ladies fainted and Robert Fitzroy (Darwin's old captain from the *Beagle*) stalked around the hall waving a Bible. On one point the legend is clear, however: Huxley had carried the day and Wilberforce was crushed.

In fact, modern historical studies have shown that the legend is a myth concocted by later Darwinians (Brooke, 2001; Jensen, 1988; Lucas, 1979; James, 2005). There is little evidence from contemporary sources that Huxley's remarks had so powerful an effect, and some who were there recorded that a speech by the botanist Joseph Hooker was more influential in saving the day for Darwin. What, then, do we make of this flawed symbol of the apparent triumph of evolutionism within a year of the *Origin*'s publication? It is clear that there was no immediate triumph of Darwinian materialism. Evolution was gradually accepted by the scientific community and eventually by most educated people in Britain and America (for surveys see Ellegård, 1958 on the popular press, and Hull, 1973 on the scientists). In America, Louis Agassiz held out for divine creation until his death in 1873. But by the 1890s Sir J. W. Dawson of Montreal was almost the only scientist of any reputation still opposing evolution (Dawson, 1890; see O'Brien, 1971; Sheets-Pyenson, 1996).

In the short term, however, the process of conversion involved much hard lobbying by Huxley, Hooker, and the other naturalists who jumped to the defense of the theory. Huxley's real triumph was in gradually extending the influence of those who shared his aversion to the design argument within the community of professional scientists. This influence was at last beginning to expand, and Huxley networked endlessly to ensure that people sympathetic to his position got the jobs that were opening up in the universities and elsewhere. It became unfashionable for a scientist to make open appeals to the supernatural, even if he (and they were still almost all men) believed in a Creator. Science was about providing naturalistic explanations of how the world works, and in the area of organic origins this meant some form of evolutionism. In that sense, Huxley triumphed over the supporters of design, even if in a less dramatic fashion than the popular image of his confrontation with Wilberforce would imply.

THE FIRST DARWINIANS

The resulting evolutionary perspective became known as Darwinism, but whether or not it was a Darwinism that we would recognize today is another matter. Natural selection found some support, but also faced numerous objections, some of which seemed more plausible then than they do today. Many of those who accepted evolution found it hard to go along with the claim that it resulted from little more than trial and error. If evolution was going to be the foundation for an ideology of social progress, it would help if something more purposeful were pushing things along. Huxley himself had doubts about the adequacy of the selection theory, and he made common cause with the social philosopher Herbert Spencer, who advocated the inevitability of progress and was committed to the Lamarckian hypothesis of the inheritance of acquired characteristics. By the standards of a modern Darwinian, then, Huxley's triumph was a slightly hollow one. He prevailed on the general question of evolution, but could not shake a popular expectation that something more purposeful than trial and error must be involved (Bowler, 1988; Moore, 1979; Ruse, 1979, 1996). The scientific case for evolutionism was built on the evidence from geographical distribution, morphology, and the fossil record. Wallace and Hooker expanded Darwin's efforts to show how the existing distribution of species could be explained by postulating expansion from centers of origin. Wallace also showed how the relationship between species and varieties could be understood in terms of local populations becoming divided by geographical barriers. The American botanist Asa Gray became one of Darwin's most active supporters, studying the distribution of North American plant species and becoming a leading advocate of a reconciliation between Darwinism and religion, as I describe in more detail later in this chapter.

Morphologists (who study the relationships between species revealed by comparative anatomy) also seized on evolutionism to throw light on the patterns Owen had once seen as evidence of a divine plan. The most active pioneer was the German biologist Ernst Haeckel, who reconstructed the history of life on earth by tracing each group back to its hypothetical common ancestor and then trying to show how the founder of each group could have emerged from a previously existing type. Haeckel coined the term "phylogeny" to denote the evolutionary history of a group, and it was he who inspired Huxley to take an interest in this line of research. Critics argued that many of Haeckel's hypothetical ancestors were too speculative to be of any scientific value. But both Haeckel and Huxley were anxious to free science from the old quasi-theological search for an underlying plan of creation. For them, the prospect of replacing Owen's idealized archetypes with real common ancestors was only too tempting.

There was some hope that the fossil record would yield hard evidence of the intermediates that the evolutionists postulated on theoretical grounds. Some of the most obvious gaps in the fossil record were indeed partially filled in (Bowler, 1996; Desmond, 1982; Rudwick, 1972). Here the weakness of the creationists' argument on the discontinuity of the fossil record was revealed. Creationists insist that every gap in the record is a genuine sign that the later forms were divinely produced. When some of the gaps are actually filled in, all the creationist can do is retreat and argue that the other gaps still remaining must be real. Although complete evolutionary sequences are seldom seen in the record, the discovery of forms bridging gaps that the creationists assumed were unbridgeable revealed the poverty of creationism as a guide to scientific research. And in the late nineteenth century enough gaps were filled in to give the evolutionists' program real momentum. Fossil horses were discovered in the American West by O. C. Marsh, linking the highly specialized modern horse back to small, five-toed ancestors— Huxley called this "demonstrative evidence of evolution." The famous *Archaeopteryx,* unearthed in Germany, showed that there were once creatures which combined characters now separated between reptiles and birds (feathers, and a mouth with teeth). The fossil did not allow a complete reconstruction of how birds evolved from reptiles, but it was an intermediate that would never have been predicted by a creationist. Later in the century a whole sequence of mammal-like reptiles was discovered in South Africa, providing confirmation of the morphologists' predictions of how some distinctive mammalian characters had evolved.

Reconstructing the history of life on earth seemed an obvious extension of the evolutionary program, but it tended to marginalize the theory of natural selection, because it was difficult to envisage the selective pressures that may have acted in the remote past. There was debate over the selection theory, but Darwin's efforts to promote it as the best explanation of how evolution worked were unavailing (Gayon, 1998; Vorzimmer, 1970). The most obvious problem centered on the roles of variation and heredity. This came to a head with a review of the *Origin* published in 1867 by the engineer Fleeming Jenkin. He pointed out that there was a weakness in Darwin's analogy between artificial and natural selection. The breeders' efforts to select for a particular character were successful up to a point, but they could never produce a distinct species because the new variety could always interbreed with the parent form. Jenkin argued that to form a new species there would have to be a saltation or a "sport of nature"—what we might call a macromutation. But even if the mutated form had a new character that gave it an advantage in the struggle for existence, it would have to interbreed with unchanged members of the original species and its superior character would be diluted in its offspring. It would be swamped by interbreeding with the unchanged mass of the population and its effect dissipated.

Historians have traditionally taken Jenkin's argument as evidence that Darwin's pregenetical view of heredity made the selection theory unconvincing at the time. His belief in what is known as "blending" heredity left his theory vulnerable to the claim that new characters would be swamped. The solution would come with Gregor Mendel's demonstration that biological characters breed true as units (and hence cannot be diluted by interbreeding)—the phenomenon that we now explain by the theory of the gene. But the selection theory could be rendered plausible even in Darwin's time. Wallace pointed out that every population shows a range of variation, with most individuals clustered around the mean (think of the variation in height in the human species—there are some very tall and very short individuals, but most of us cluster around the average height). There is no need for discrete new characters. The geneticists would later have to reconcile their theory of unit characters with the continuous range of variation observed in most populations.

The fact remains that many of Darwin's contemporaries felt uncomfortable with the selection theory. This was in part because of the technical issues highlighted by Jenkin, but there was also wide-spread suspicion of the idea that a mechanism based on random variation, a process of mere trial and error, could reproduce the purposeful structures traditionally attributed to design. This was Wilberforce's point in his debate against Huxley, and he was by no means the only person to feel this way at the time. The astrono-mer Sir John Herschel dismissed natural selection as the "law of higgledy-piggledy." Richard Owen and his disciple St. George Jack-son Mivart also argued that some more purposeful force must be involved (Gruber, 1960; Rupke, 1994). Mivart's *Genesis of Species* of 1870 provided a mass of anti-selectionist arguments, many of which are still used by creationists today.

The arguments in favor of purpose in nature were increasingly articulated not through creationism but through the assumption that evolution had a more positive driving force than Darwin imag-ined (Bowler, 1988). Evolution was taken for granted—it was the Darwinian explanation of it that was failing to take hold. Selec-tion might play a negative role in eliminating the less successful of nature's products, but it was incapable of playing the creative role that Darwin had assigned to it. For many religious thinkers, this meant that the divine purpose must be built into the very laws of nature which drove evolution toward its goals (Livingstone, 1987; Moore, 1979). This is theistic evolutionism, which seeks to combine creationism's commitment to the argument from design with the evolutionists' insistence that nature is governed by law rather than miracle.

In America, Asa Gray—a devout Presbyterian as well as a scien-tist—tried to defend Darwinism against the charge that it was necessarily atheistic (Dupree, 1959; on Darwinism in America see Daniels, 1968; Loewenberg, 1969; Numbers, 1998; Pfeifer, 1974; Roberts, 1988, and Russett, 1976). In the essays reprinted in his *Darwiniana* of 1876 Gray argued at first that it didn't matter what mechanism God chose to achieve His ends, as long as a beneficial adaptation was produced. But he too was worried about the many useless variations, the "scum of nature," that had to be eliminated for selection to work. In the end he conceded that God had somehow loaded the dice so that variation always produced useful characters. He advised Darwin to assume that "variation has been led along certain beneficial lines" (Gray, 1876: 148), prompting Darwin to respond that if this were so, selection would be unnecessary.

These examples show us that the distinction between Darwinian and anti-Darwinian evolutionists could often be a matter of emphasis and rhetoric rather than of theoretical division. Scientists joined one camp or the other in part because of their personal or professional loyalties. The theological divisions were similarly complex, so that it is often difficult to understand why a religious thinker was for or against evolution without knowing the local circumstances in which they operated. Huxley led a campaign to professionalize biology by eliminating references to design—yet Gray was a professional scientist too, and he was openly campaigning to supplement natural selection with a divinely purposeful force. Many liberal clergymen endorsed evolutionism. Frederick Temple, a future archbishop of Canterbury, preached a sermon during the 1860 Oxford meeting of the British Association criticizing the church's traditional reliance on apparent gaps in the world's history as evidence of divine intervention. Temple's *The Relations between Religion and Science* of 1884 openly supported evolution as the mechanism of creation. Another Anglican clergyman, Charles Kingsley, best known as the author of *The Water Babies*, also saw no reason why God's purpose should not be worked out by law rather than by miracle.

The critical question was whether or not evolution showed evidence that the laws of nature worked toward a preconceived goal. Owen and Mivart opposed Darwin because they saw the development of life as something based on a coherent divine plan, not an endless sequence of local adaptations. They expected that plan to be built into the very structure of the world—it could not be an indirect trend produced by the summing up of a host of apparently unplanned interactions. It is doubtful that Huxley himself would have agreed with Darwin that evolution was nothing more than a mass of individual adaptive responses in each separate line of evolution. He was not opposed to Owen's view that something imposed a degree of unity on the development of life. But where Owen wanted to see this "something" as the expression of a divine plan, Huxley preferred to think in terms of natural forces limiting the pathways available to evolution. Variation was not random, as Darwin supposed—it was directed along certain lines, although by purely natural forces.

There were thus real tensions within the Darwinian camp between those who followed Darwin in seeing evolution as purely adaptive, and those who thought that biological constraints pushed evolution along restricted pathways. This explains why there was a somewhat uneasy alliance between Huxley and the most influential philosopher of the evolutionary movement, Herbert Spencer (Peel, 1971). For Spencer, long-range trends were never "locked in"—they were always the summation of a host of individual decisions and actions. His model was the free-enterprise capitalist system, which was the key to social progress because no government could plan in a way that would successfully integrate all the factors involved. The only way forward was to let nature take its course, which meant allowing individuals to make their own decisions and sink or swim by the consequences. Struggle and competition were agents of progress because they stimulated everyone to improve themselves and to adapt to changes in their environment—and Spencer was convinced that the results of self-improvement could be passed on from parent to offspring. He was a Lamarckian evolutionist long before Darwin published, and although he coined the term "survival of the fittest" to denote the operations of natural selection, he always thought that self-development was more important than selection. Huxley and Spencer made common cause to promote the interests of the new elites against the old conservative social program based on aristocracy and the ownership of land, but Spencer's elite were the industrial entrepreneurs, whereas Huxley's were the professional managers and scientists. Spencer saw evolutionary progress as inevitable—but it was inevitable because the forces of nature always favored the fittest in the long run, not because the direction was built in by nature, let alone by God.

If Huxley and Spencer shared their opposition to a science in which nature and human nature were seen to be divine artifacts, they also shared the expectation that nature was a progressive system which would, in the end, work its way up to something like the human form. There was no explicit design in their systems, because neither believed in a supernatural agency controlling nature. Yet the assumption that evolution is inherently progressive allowed them to retain a sense of purpose in nature which could look remarkably similar to the natural theologians' argument from design. Small wonder, then, that as Moore (1979, 1985a and b) points out, many liberal clergymen were willing to see themselves as followers of Spencer and hence as Darwinians. Here we come to a central feature of nineteenth-century Darwinism that must seem hugely paradoxical to a modern American creationist. Religious thinkers welcomed evolutionism as long as they could see it as a purposeful process—exactly what Spencer and the agnostics were offering them. It was a small step from Asa Gray's hope that God steered variation in the right direction to Spencer's claim that the individual's response to its environment was the driving force of Lamarckian evolution. And because Spencer's Lamarckism stressed the role of struggle and competition, it was hailed as a form of Darwinism, even though it subordinated natural selection to self-improvement. Like Darwin, Spencer saw progressive evolution as the summing up of a vast number of individual interactions. Spencer's social policy would thus become known as the classic form of social Darwinism in the later nineteenth century (Hofstadter, 1959; Hawkins, 1997). Yet unlike the pure selection theory, it could be endorsed by liberal religious thinkers because it embodied the traditional Protestant virtues of hard work, industry, and initiative as the driving forces of progress (Livingstone, 1987; Roberts, 1988).

We can see this liberal Christian response to Darwinism in a book familiar (if only in a cartoon version) to almost everyone, *The Water Babies,* written by Charles Kingsley in 1862. We have already noted Kingsley as one of the liberal Anglican clergymen who welcomed Darwin, and his children's story of Tom, the chimney sweep's boy who becomes a water baby, offers a wonderful commentary on the social values that drew people to Spencer's Lamarckian version of Darwinism. Huxley and Owen are introduced as scientists who debate whether or not water babies are contrary to nature—with Kingsley warning them that if they dare to be so dogmatic, the Queen of the Fairies is likely to astonish them. Creation here is done not by God himself, but by supernatural agents to whom He has delegated His powers, and these powers never work by direct intervention. They do their job by putting pressure on living things to encourage their self-development. When Tom encounters Mother Cary, the personification of nature who makes new animals all the time, he is astonished to find her sitting still rather than engaged in a frantic process of manufacturing each new type. She doesn't need to do this, she tells him, because "I sit here and make them make themselves" (Kingsley 1889: 273). Paley's industrious clockmaker Designer has been replaced by the creative forces of nature, operating behind the scenes to ensure that the good are rewarded and the bad punished. This is a harsh world in which you have to work hard and make sacrifices to succeed, but the rewards are personal development and a contribution toward the progress of society. If you take things easy you will degenerate, as in the story of the Doasyoulikes who eventually become apes (1889: 229–237). The moral, proclaimed to our animal ancestors at the end of the book, is that if you work hard and keep yourself clean you will eventually become human.

It is easy to see why this might be taken as Darwinism: evolution has replaced supernatural miracle in the origin of species, and it works through a harsh mechanism that is unforgiving to those who do not respond to nature's challenges in a productive way. But this is pure Spencerian Lamarckism, in which struggle stimulates self-improvement for the individual and hence for the race. It appealed to liberal Christians like Kingsley because it encapsulated the virtues of the Protestant work ethic in the actions of nature herself, thereby confirming that the world of living things was created by a God who intended them to "make themselves." Only by transferring the human values that were once thought to lift us above nature into nature herself was it possible for a liberal Christian to see evolution as a force that could have created the human mind as well as the body. Conservatives like Wilberforce rejected this effort to see nature as creative, and probably had a better appreciation of the true implications of natural selection. In fact, some of the liberal Anglicans found the evolutionary origin of the human spirit a hard pill to swallow, including Frederick Temple and Aubrey Moore. It was all very well to see evolution as a form of design, but if you wanted to see humanity as being in need of a savior, then there had to be something that separated us from the march of progress.

It was in America where the Spencerian form of self-help evolutionism made most headway among liberal clergymen—just as it was here that his form of social Darwinism became the public expression of capitalist values. James Moore (1985a) notes that when Spencer was given a grand dinner at Delmonico's in New York in 1882, there were clergymen present who had hitched their wagons to the star of his hugely popular philosophy. It was the robber barons of American industry who staged the event, and for them it was important to celebrate the work of a philosopher who endorsed their ideology of progress through competition. But the clergy too had seen—as had Kingsley in Britain—that a philosophy which built the Protestant work ethic into nature was a valid foundation for the religion that they used to give a veneer of respectability to what their political opponents would label as a validation of brute force and selfishness. The philosopher John Fiske—raised a Congregationalist and now a leading religious liberal—had led the way in adapting Spencer for a religious audience. His *Outline of Cosmic Philosophy* of 1874 had stressed how natural evolution led up toward the higher values expressed in the human mind. At Delmonico's he responded to the toast to Spencer with the words: "Evolution and religion: that which perfects humanity cannot destroy religion."

Also present were the Congregationalist clergymen M. J. Savage and Henry Ward Beecher. Savage, minister of the Church of Unity in Boston, had written books on the theme of evolution as a moral process, *The Religion of Evolution* of 1876 and *The Morals of Evolution* of 1880. Beecher preached at Plymouth Congregational Church in Brooklyn and was one of the most articulate of the new liberal theologians. Here, in another response to the toast to Spencer, he hailed the British philosopher as a guiding light who had shown him how to throw off the shackles of Calvinism and the doctrine of Original Sin. In a collection of sermons published a few years later under the title *Evolution and Religion,* Beecher developed these points, making clear his rejection of the idea of the Fall.

> And in that sublime Apocalyptic Drama, in which the struggle between good and evil, organized into laws, governments, and institutions, is carried forward with sublime mystical treatment to the final victory of good, no place is found for Adam, and no place for any illusions, even, to the malformed and monstrous doctrine of the fall of the race of Adam, and its alleged terrific consequences, which have become the bedrock on which theology has been built. (Beecher, 1885: I, 92)

Significantly, Beecher argues the case by stressing that it is in Christ's words, as reported by the New Testament, that there is no mention of the Fall. Here is the very point that would raise howls of anguish among British Christians forty years later when made by Bishop Barnes in his "gorilla sermons." Yet it was made by an American clergyman who was merely expressing his acceptance of Spencerian evolutionism. Progress was the key to recognizing that the values of American capitalism offered the way to the perfection of humanity in this world rather than the next. The Bible could be reinterpreted in the light of Darwin and Spencer—and of their disciples, Andrew Carnegie and the other American capitalists who were there at the 1882 dinner to celebrate their idol.

HUMAN ORIGINS

Social Darwinism was a manifestation of liberal values acceptable both to the radicals who campaigned to overthrow traditional religion and to more progressive religious thinkers. The conservative position was identified not by opposition to evolution within the animal kingdom, but by a reluctance to accept that general evolution—even if divinely superintended—could generate the spiritually unique characters of the human soul. Evolution worked as an ideology of progress only for those who were willing to accept that the progress seen in animal evolution was aimed at the production of the human mind, and hence that social progress was a continuation in another sphere of the basic processes of evolution. For this reason the debate over human origins was a crucial aspect of the controversy sparked by Darwin's theory, and the overall success of evolutionism has to be measured by the temporary retreat of the conservatives in the face of a growing willingness of many ordinary people to believe that humans were, after all, the products of a natural and not a supernatural process. This move was only made possible, though, by a course of action that evaded the more materialistic aspects of Darwin's theory and allowed evolution to be presented as a process with an overall purpose. Spencerian self-improvement was always more attractive than the random variation of biologically fixed characters followed by a ruthless winnowing out of those unlucky enough to be born congenitally unfit.

The Darwinians' efforts to make the case for human evolution came at an appropriate time. The previous decades had seen a major transformation in people's willingness to accept that the human mind was in some respects a manifestation of physical processes going on in the brain. This in turn made it easier to see the human mind as a product of natural evolution—not a purely spiritual entity somehow plugged into the body from a supernatural dimension. In another area entirely, the Darwinian debate coincided with a profound revolution in ideas about human antiquity, a revolution that opened up the vista of a long, prehistoric past during which our ancestors had raised themselves slowly from a state of primitive savagery. This in turn made it possible to believe that these earliest ancestors had not been fully human, but had still retained some aspects of their origin in the apes.

The new ideas on the nature of the mind owed a great deal to the science of phrenology, already noted in the previous chapter (Cooter, 1985). Phrenology was an attempt to show that the faculties of the human mind were directly linked to corresponding areas of the brain. According to the phrenologists a trained observer could tell someone's personality from the shape of their brain—each mental faculty was produced by a particular part of the brain, and if that area of the brain was well-developed, then so was the corresponding personality trait or faculty. Phrenologists believed that the skull reflected the shape of the brain, and hence claimed to be able to read the personality from the "bumps" on the skull. For this reason they were widely dismissed as charlatans by the scientific establishment, but the underlying philosophy of phrenology had profound implications. It seemed to endorse the materialists' claim that the mind was a by-product of the brain. And while the professional scientists sneered, the general public lapped up the ideology of the new "science" of the mind and thus became used to the idea that the mind was the product of nervous activity, not of a purely spiritual soul which could be unplugged from the body at death.

If mental activity was generated by the brain, it was a fairly short step to argue that if the brain became larger and more complex in the course of evolution, then the mental powers displayed by animals would also increase in sophistication. Chambers took this step quite explicitly in his *Vestiges* of 1844, boldly proclaiming that the human mind was a product of the increased size and complexity of the brain. Herbert Spencer, himself strongly influenced by phrenology as a young man, took things one step further in his *Principles of Psychology* of 1855. Here he applied a Lamarckian model of evolution to the mind, suggesting that as individuals developed their own skills and mental powers, the results were passed on to future generations and defined the inherited capacities and instincts of the race. The Darwinians would not be the first to argue the case for the evolutionary origin of human mental powers, although the debate certainly became more active after Darwin published (Richards, 1987).

The revolution in ideas about human antiquity came about far more suddenly, and at just the right time to aid the Darwinians' case. Geologists had established that the earth itself was immensely old, but in the early nineteenth century it was still believed that the biblical chronology remained valid for the human race. There were no human fossils, and it was still possible to believe that the first humans had been created as in the Genesis account, only a few thousand years ago. There was no prehistory for the human race, and our earliest ancestors had been civilized from the start—the great empires of Egypt and Babylonia went back almost to the time of Noah. When geologists and archaeologists began to turn up

stone tools and human remains from geological formations contemporary with the mammoth and other extinct animals of the ice
age, they were dismissed as later intrusions. Only in the late 1850s
did a group of British geologists begin to take these discoveries seriously. They traveled to France to see the chipped stone tools being
unearthed by Jacques Boucher des Perthes from the gravel beds of
the Somme valley, and became convinced that they were genuine
(Grayson, 1983; Van Riper, 1993). Here was clear evidence that humans with a very primitive level of technology had existed at a date
that was many thousands of years earlier than any of the events recorded in the Bible. In 1863 Charles Lyell summed up the new con-
sensus in his *Geological Evidences of the Antiquity of Man.* The discovery of human antiquity did not by itself imply that our earliest
ancestors had evolved from apes—Lyell himself found this hard to
believe—but it removed one major argument against the Darwinian position. There were still no fossil human remains, but it was
plausible to believe that the creatures who made these primitive
stone tools were not fully human.

Given these developments, it was inevitable that the debate over
the *Origin of Species* would be extended to include the origin of the
human race. Darwin tried to steer clear of this tricky subject by
mentioning it in only a single sentence. But everyone knew what
was implied, and that was why Wilberforce was so easily able to include a gibe about Huxley's ape ancestry in the 1860 debate at the
British Association. In fact, Huxley was already involved in a debate
with his rival, Richard Owen, over the closeness of the relationship
between humans and apes. Owen wanted a clear separation based
on the claim that there was a structure in the human brain, the hip-
pocampus minor, that was not present in the apes. Huxley argued
that there was no such difference: Owen was an incompetent anatomist biased by his religious preconceptions, and humans and apes
should be classified in the same order, the Primates. In *The Water
Babies,* Charles Kingsley lampooned them both by presenting this

as a debate over the presence of a hippopotamus in the human brain (1889: 153). But there was a serious point to Kingsley's ridicule, since he noted that the real difference everyone cared about was our ability to tell right from wrong and to say our prayers. Owen had given the game away by accepting the phrenologists' point that the brain was the organ of the mind. If the mind were a product of a spiritual entity, the soul, why should it need new physical organs through which to function? Even if the human brain is different to that of the ape, the materialism inherent in the evolutionists' position has already been conceded: nothing genuinely new is needed to explain the higher mental faculties we possess. If the anatomical differences were indeed only trivial, as Huxley asserted, then the inclusion of humans into the world of material nature was complete.

This point was made in the title of the 1863 book Huxley published to sum up his contributions to the debate with Owen: *Man's Place in Nature.* At the physical level, there was nothing significant to separate humans from apes, and by implication there could be nothing at the mental level either. Humans have bigger brains, and as one might expect, they have higher mental functions, although Huxley said nothing about how those higher functions might have been produced. He did make it clear elsewhere that he was a determinist, although not (he insisted) a materialist. Mental functions are by-products of physical activity in the brain, and as such are rigidly predetermined by natural law. Any feeling we have of "free will" is an illusion. This was a key aspect of the position sometimes called "scientific naturalism"—indicating that it left no room for entities or actions lying outside the world of natural law. In 1874 Huxley's friend, the physicist John Tyndall, gave a notorious address at a British Association meeting in Belfast in which he insisted on science's ability to explain all the phenomena once labeled as supernatural. Here the assumption that humans had evolved from apes was merely part of a much wider campaign to replace religion with science in the study of all aspects of nature, including human behavior.

The one thing that was lacking to make the evolutionists' case complete was the "missing link" (as it later came to be called), a fossil intermediate between the ancestral ape and a modern human. Such a creature would by definition have a brain somewhere in between the size of an ape's and of a human's. Huxley looked at the available specimens of ancient humanity, in particular the skeletal remains unearthed at Neanderthal in Germany in 1857. Here was a creature with a very ape-like appearance, the skull showing heavy brow-ridges and a very sloping forehead. The temptation to label the Neanderthal specimen a missing link must have been very strong, but Huxley was honest enough to accept that the brain capacity was as large as that of a modern human. Although some later authorities would accept the Neanderthals as intermediates between humans and apes, Huxley conceded that they could not serve as a convincing link. For the time being, the evidence for human evolution from the apes remained circumstantial (see Bowler, 1986; Reader, 1981).

Most of the early evolutionists assumed that the expansion of the brain—and hence of intelligence—was the main driving force of progressive evolution leading toward humanity. Lamarckian use-inheritance and natural selection would both tend to enhance intelligence. Increased intelligence was both a consequence of individuals developing their mental powers during their own lifetime, and a primary survival factor favored by selection. If the successive stages in this expansion were not revealed by the fossil record, some clues as to the later stages in the sequence could be found by looking at what were widely regarded as the "savage" forms of humanity surviving in some parts of the world. The missing link was not truly missing—it could be seen in places like Australia where the aborigines were still living in the stone age when discovered by

Europeans. Most physical anthropologists also believed that these "lower" or "primitive" races had smaller brains than Europeans (Gould, 1981).

Evolutionism thus provided arguments used to defend the race theories of the later nineteenth century (Haller, 1975; Stepan, 1982). Those races conquered or exploited by the Europeans were regarded as relics of the past, retaining characters now surpassed by the triumphant whites. Modern creationists often argue that Darwinism played a major role in the emergence of racism, forgetting that the creationists of the nineteenth century also endorsed white superiority. Louis Agassiz, who became the founding father of American biology, remained a committed creationist until his death in 1873 (Lurie, 1960). Yet he supported the arguments of slave-owners by insisting that blacks were a separately created species destined to remain subordinate to the whites. Both creationism and evolutionism could be used to provide support for the increasingly popular ideology of white supremacy.

By the time Darwin published his own *Descent of Man* in 1871 he was contributing to a debate that was well underway. In this book he drew on the work of Huxley and others for the evidence suggesting a close link between humans and apes. He also accepted the figures quoted by physical anthropologists who claimed that races such as the Australian aborigines had smaller brains than Europeans. Darwin tried to minimize the differences between animal and human mental powers so he could blunt the force of the claim that here was a gap so wide that natural evolution could never bridge it. He was thus accepted the claim that some of the most "savage" tribes had limited mental faculties. He also accepted a great deal of anecdotal evidence from observers such as zookeepers and hunters about the mental agility of animals. He was even prepared to admit a strong element of altruism in animals, as when a monkey defended a zookeeper who had been attacked by a fierce baboon, at the risk of its own life. Modern psychologists regard much of this evidence as unreliable, prompted by an all-too-common urge to anthropomorphize animal behavior.

Darwin stressed that at both the physical and the mental levels, humans were more closely related to the apes than was acknowledged in the traditional viewpoint. There was no qualitative difference, as would be expected if we had a spiritual faculty not possessed by the animals—only a difference of degree. But he was more concerned than Huxley to explain precisely how evolution had changed the ancestral ape into a bigger-brained human. Like Spencer, he accepted that the human mind (and hence the brain) had been shaped and developed by forces which were ultimately driven by survival value. Lamarckism could account for how new faculties were added by the efforts of successive generations to deal with the challenges posed by their environment (including the social environment). Natural selection too could be expected to promote increased intelligence as a survival factor.

In one respect, though, Darwin went beyond the assumptions of his contemporaries. He saw that the separation of the human and ape families had to be explained in terms of adaptive divergence. If intelligence was such a good thing, why hadn't the apes continued to develop it at the same rate as our own ancestors? The conventional view supposed that both the apes and "primitive" humans had lagged behind because they were not exposed to the same level of environmental challenge. The tropics were seen as a "softer" environment than the more northern climates—which is why many early evolutionists refused to accept Darwin's view that the human species had evolved in Africa. Darwin realized that it was necessary to specify an adaptive shift which would explain why the ancestors of humans had been exposed to different selective pressures from those experienced by the remaining apes. He argued that our forebears had moved out of the trees onto the open plains, and had stood upright because this offered a more efficient means of loco-motion in this environment. The hand had then been perfected for manipulating sticks and stones, and the drive to make improved tools might explain why our intelligence developed to a higher level. Unlike almost all of his contemporaries, Darwin thought the earliest hominids had walked upright before they began to get big-ger brains—a striking anticipation of the view confirmed by later fossil discoveries.

For most religious believers, however, it was the moral power of the human mind that defined our existence in a spiritual world not shared with the animals. Here Darwin faced his greatest challenge, that of explaining our conscience and our willingness to sacrifice ourselves for others, in terms of natural evolution. To many, it seemed absurd that natural selection could promote the instinct for self-sacrifice—surely anyone developing such a character by ran-dom variation would be eliminated in the struggle for existence. Natural selection would promote selfishness, not altruism. But Her-bert Spencer had already ventured into this realm and had begun the process of explaining how evolution might produce cooperative behavior. Once we accept that our ancestors—like some of the great apes—lived in social groups, it becomes easier to see how evolution might temper selfishness with instincts designed to promote coop-eration within the group. If membership of the group offers advan-tages, then individuals who cut themselves off by too selfish behav-ior will be expelled and will lose those benefits.

As a Lamarckian, Spencer accepted that cooperative habits would eventually become inherited instincts. Darwin himself acknowl-edged a role for Lamarckism in this area, but he also appealed to a process now known as group selection. If a species is divided into competing tribal groups, then a group whose members cooper-ate effectively will displace one whose members do not pull to-gether. Selection will then promote the instinct to sacrifice one's own interests for those of the group. For Darwin, our so-called moral sense is merely an intellectual awareness that our behavior is conditioned by these implanted cooperative and altruistic instincts. This explains why in practice it is always easier to persuade people to sacrifice themselves for those they are familiar with than for total strangers. Only by intellectual abstraction do we arrive at universal moral values that are supposed to be applied to all—and many of us find it very hard to live up to such ideals.

If Huxley persuaded people that they were related to the apes, it was Spencer and Darwin who provided the arguments that would encourage them to believe that human nature itself is a product of the natural world. They showed how evolution could develop the characters we recognize as essential for morality. This challenged the traditional Christian view that human nature has mental and moral components derived from an immortal soul. Coupled with Spencer's efforts to show that evolution promoted the values of the Protestant work ethic, these moves made it possible for the Darwinists to present their theory as a modernization of traditional Christian culture rather than a complete negation of it. For every agnostic such as Huxley, there was a liberal Christian who welcomed evolutionism because it built the highest characters of the human mind into nature and made them the driving force of progressive evolution. The emergence of the human mind from the mentality of the higher animals merely completed the process of progressive evolution. The image of natural selection as a meaningless cycle of suffering that could promote only selfishness and brutality was sidelined. God had created a moral universe and had given it the power to build the human mind from primitive beginnings. Many liberals believed that our actions in the cause of personal development and social progress are continuing the divinely instituted process that created us, and that will lead to a future humanity in which all those higher elements are perfected.

A STEP TOO FAR?

In an age fascinated by the progress achieved in the industrial revolution and determined to spread the values of Western civilization around the world, the Spencerian vision of evolution found a ready audience. The liberal Christians joined with the secularists in accepting that the perfection of humankind was to be achieved in this world, not the next. The only difference was that the liberal Christians saw progress as a vision of God's ultimate purpose, not just a purely natural trend. To make this move they had to modify the traditional Christian view that human nature is fundamentally depraved and can only be perfected in the next world. Talk of Original Sin and of the Fall of Man plays a much smaller role in the theological writing of this period. Yet even liberal Christians found some incongruity in the effort to maintain that humans are merely improved animals. To retain any continuity with the traditions of their faith, they had to believe that humans were capable of sin, and to give the concept of sin any meaning, human nature had to be something more than the mentality of a social animal. Some of the liberal Anglicans who accepted evolution as the unfolding of a divine purpose nevertheless found an evolutionary explanation of the soul unconvincing. Frederick Temple could accept the evolution of the human body, but conceded that the appearance of the soul required an act of divine creation, and Aubrey Moore acknowledged this as the most difficult area of the whole controversy (Moore, 1889: 92, 200–215).

To conservative theologians such as Wilberforce, the whole idea of a natural origin for human nature was unthinkable. By highlighting the brutality of natural selection they hoped to show that our spiritual capacities could never have evolved by such a process. The scientists who opposed Darwin shared this viewpoint. Richard Owen and St. George Mivart both became theistic evolutionists, but neither could accept that human nature was produced gradually out of the ape's mentality. Owen preferred to argue the case in terms of anatomical differences in the brain, but Mivart—a devout Roman Catholic—argued against the *Descent of Man* on moral and religious grounds (his review is reprinted in Mivart, 1892: II, chap. 1). He thereby earned the enmity of Huxley and the agnostic Darwinians, but he was articulating what would eventually become the orthodox position of his Church. (We shall see in the next chapter that this did not come about until the twentieth century. Mivart was excommunicated just before his death; see Gruber, 1960).

Some of Darwin's scientific supporters also found the extension of the theory to include human nature distasteful. Charles Lyell, the apostle of continuity in the physical world, had always been troubled by the idea that humans should be included within the natural system (Bartholomew, 1973). His long resistance to the basic idea of evolution was almost certainly fueled by his fear that it would be applied to human origins. In his *Antiquity of Man* he provided lukewarm support for Darwin's theory of evolution, but explicitly argued that there would be a sudden leap (presumably of supernatural origin) at the creation of the first true humans. Darwin wrote that this remark made him "groan," but worse was to come because even Alfred Russel Wallace was beginning to doubt the ability of natural evolution to produce the human mind. As the co-discoverer of natural selection, Wallace became one of the theory's most consistent supporters during the 1860s, less willing than Darwin to accept a role for other mechanisms of evolution. Yet by the end of the decade he began to argue that many aspects of the human mind can have conferred no selective advantage and thus cannot have been developed by evolution. Unlike many of his contemporaries, Wallace did not share the view that "savages" were mentally inferior to whites—but he pointed out that in their way of life (presumably equivalent to that of the earliest humans) the higher mental functions played very little role. How can those functions have been developed at a time when all humans shared such a primitive life-style? Wallace finally came out openly in favor of the view that some form of supernatural guidance had shaped the later stages of human evolution. Significantly, he had just become convinced of the reality of spiritualist phenomena, and it seems that acceptance of a soul which can survive the death of the body played a major role in his change of heart on this issue (Kottler, 1974; Fichman, 2004; Turner, 1974).

Wallace's refusal to follow Darwin and Huxley into a completely naturalistic worldview makes a convenient point at which to bring our survey of the Darwinian revolution to an end. It highlights the difficulties arising from any attempt to present the debate as a simple battle between evolutionary materialism and traditional Christianity. Wallace was a religious man who apparently saw no incongruity in the idea that evolution, even if driven solely by natural selection, could be accepted as God's mechanism of creation. Yet he could not follow Darwin, Huxley, and Spencer into a completely naturalistic account of human nature. Wallace remained on good terms with the Darwinians, while Mivart was ostracized by the group for expressing views that were in some respects very similar. The difference was that Mivart campaigned actively against any purely natural explanation of evolution, and did so as a member of a Church to which Huxley felt particular antipathy. His willingness to see evolution as the unfolding of a divine plan was increasingly unacceptable to the scientific community, even though there were many scientists who felt that the system of natural laws which governs the world (and hence produces evolution) was of divine origin. Huxley's point was that the divine plan was not an active directing agent which could be detected by science—even though he himself doubted that variation was truly random as Darwin had supposed.

By the 1870s support for outright creationism (as we would call it today) had been marginalized within the scientific community, although it would be wrong to imagine that most scientists were materialists and agnostics. Many educated laypersons had also become willing to accept some form of evolutionism and might have called themselves Darwinists. But both in science and in the wider community, most people wanted to feel that evolution offered something more purposeful than the most materialistic reading of Darwin's theory would imply. They wanted evolution as a mechanism of progress that laid the foundations for human progress in this world. For agnostics such as Huxley and Spencer, progress was simply a natural consequence of the laws that science was discovering. Humans had only to recognize and apply those laws in everyday life to ensure the future perfection of humanity. But since Spencer had shown how the values of the Protestant work ethic could be seen underlying the laws of nature, liberal Christians could also take this progressionism on board. Some still felt that evolution had to be seen as the unfolding of a divine plan, but others were prepared to accept that only the most basic laws of nature flowed from the Creator—all that was produced by their interactions was part of His intentions. Emphasis on Original Sin and the Fall diminished, although many religious thinkers still found it hard to accept that our spiritual faculties could be explained away by natural evolution from the animals.

In one sense, then, Darwinism had triumphed, but tensions remained that would define the ongoing debates of the next century or more. Conservative Christians such as Wilberforce had highlighted the materialistic character of natural selection, and they would have nothing to do with Spencer's Lamarckian modification of the system. Problems were bound to emerge if the Lamarckian alternative were discredited in science, leaving everyone confronted with the full implications of selectionism. Spencer's vision of struggle as the spur to self-improvement was a reflection of the individualist ideology of nineteenth-century capitalism, and as alternatives to that ideology were explored, the logic of his modified social Darwinism would seem less compelling. Some of these alternatives would share the conservative Christians' dislike of natural selection, although for very different reasons. But from the viewpoint of the debate between science and religion, the most significant issue that was left outstanding was the extent to which the ideology of progress and evolution required a transformation—or a betrayal—of traditional Christian values. Sooner or later, the liberal Christians were going to have to face up to the fact that by accepting the possibility of future progress, they had abandoned the idea of Original Sin and would have to find a very different meaning for Christ's sacrifice on the cross. Conservatives, who had seen the difficulties from the start and had always stressed the materialistic implications of evolution, would eventually rally their forces and renew their assault on Darwinism.

THE ECLIPSE OF DARWINISM

At first sight, it might seem that the outburst of fundamentalist opposition to evolutionism that led to the Monkey Trial of John Thomas Scopes in 1925 interrupted a period of relative calm in the debate. By the 1870s Darwinism had been widely accepted even by many religious thinkers. The more liberal approach to Christian theology had sanctioned the belief that evolution developed according to a divine plan. In science, the very restricted acceptance of Darwin's theory of natural selection allowed more purposeful mechanisms of evolution to play a significant role. From a superficial viewpoint, this situation seems to have remained largely unchanged through into the early twentieth century, when the rise of fundamentalism at last galvanized resistance to the evolutionary paradigm. By 1925 this new element had led to the banning of evolutionism from some American schools and hence to the Monkey Trial.

Deeper historical analysis reveals that the decades separating the original acceptance of Darwinism from the Monkey Trial were by no means devoid of incident. Indeed, by 1925 the situation in both science and society had changed dramatically. The loosely defined Darwinism of the 1860s and 1870s collapsed as alternatives such as the Lamarckian theory were refined as complete alternatives to natural selection. Julian Huxley, one of the founders of the modern selection theory, looked back on this period as an "eclipse of Darwinism." But genetics soon began to undermine the Lamarckian alternative, and in the 1920s a synthesis between genetics and the selection theory began to emerge. By 1930 it was becoming more difficult for scientists to pretend that natural selection played only a minor role in the evolutionary process. The debates of the later twentieth century would be all the more strident because Darwinism had been tried in the fire and had emerged in a purer form, with its radical implications in full view.

At the same time, religious thinkers reflected on the situation created by the liberals' eager acceptance of Spencerian evolutionism during the 1870s. By the end of the century there was increasing recognition that the original form of Darwinism, for all that it limited the influence of the selection theory, was still very closely tied into a mechanistic vision of the world in which everything was driven by individual selfishness. To some extent, the "eclipse of Darwinism" in biology was inspired by the search for an alternative vision that would incorporate non-Darwinian mechanisms into a worldview transcending Spencer's free-enterprise ideology. In the early decades of the twentieth century, enthusiasm for Henri Bergson's philosophy of "creative evolution" led to the whole edifice of Darwinism being dismissed as a product of soulless materialism. Evolution had to reflect something more purposeful than individual selfishness, either the unfolding of a preordained plan, or the upward struggle of a purposeful life force. The emergence of an explicitly anti-selectionist evolutionism represents the first major challenge to Darwinism from a religious or moral perspective (apart, of course, from early opponents such as Wilberforce).

Inspired by the opposition to selectionism in science, the theological liberals of the early twentieth century renewed their efforts to create a synthesis with evolutionism—but now it was with the very non-Darwinian version of evolution promoted during the "eclipse." It was in the context of this revival of interest in

evolutionism that Bishop E. W. Barnes delivered his "gorilla sermons." A few liberals may have realized that accepting evolutionism
meant abandoning the idea of Original Sin in order to accomodate
the ideology of progress. But Barnes felt that many Christians had
not thought through the implications of this step—indeed many
had failed to recognize that it undermined most of the traditional
foundations of their religion. Deep issues had been skated over in
order to avoid seeming to be out of step with modern thought.
Barnes was associated with a group called the Modernists within
the Anglican Church, and he was determined to free Christianity
from its dependence on a number of ancient superstitions. Exploring the implication that we had risen from the apes was only
one part of this program. But to many Christians, his strident calls
for reform only highlighted the danger that the whole purpose
of the faith would be lost amid the call to improve humankind in
this world. Christ would become merely a great teacher, not the
Son of God who suffered on the cross to save us from damnation.
In America the fundamentalists articulated these issues and demanded that the compromise with evolutionism be rejected. In
Europe there was no surge of evangelical fervor in the early twentieth century—but Barnes would soon see his Modernism undermined by a rise of neo-orthodoxy as Europe drifted toward the cri-
sis of war.

Curiously, Barnes was one of the few theologians who realized
that the situation in biology was changing too, and that in future
religious thinkers would have to deal with a resurgent Darwinism.
He, at least, sensed that the compromise based on liberal theology and non-Darwinian evolutionism was under threat from both
sides. In 1925 all eyes might have been fixed on Dayton, Tennessee,
for the trial of John Thomas Scopes, but changes were taking place
in science that would propel the debate into a new phase. It seems
unlikely that those assembled for the trial had fully appreciated the
nature of the coming conflict. Most of the scientists the defense

would have liked to have called as witnesses were senior figures unaware of the new Darwinism emerging in biology. In the churches, the battle between fundamentalism and the liberal vision of progressive evolution was obvious to all—but few would have anticipated the extent to which that vision would be threatened by developments taking place in both science and society at large.

FROM NEO-LAMARCKISM TO CREATIVE EVOLUTION

Because the theories explored during the eclipse of Darwinism are now discredited, the whole event was ignored by historians until comparatively recently (for my own accounts see Bowler, 1983, 1988, 1996). We now recognize that the first generation of Darwinians were not very Darwinian by modern standards, since they accepted only a limited role for natural selection. But equally significant is the wave of explicitly anti-Darwinian feeling that swept through science at the end of the nineteenth century, providing a clear line of demarcation between the original form of Darwinism and its far more narrowly defined modern namesake. This was not opposition to evolutionism itself—virtually all educated people at the time accepted the basic idea that life had developed on earth by some kind of natural process. But now there was a definite move to replace Darwinism with alternatives uncontaminated by that theory's reputation for materialism. It was only when genetics destroyed these alternatives that the selection theory could emerge as the dominant mechanism of evolution.

The most prominent alternative to selection was the Lamarckian theory of the inheritance of acquired characteristics. This seems paradoxical because Lamarckism was an integral part of Spencer's theory of evolution, and was accepted as a subsidiary mechanism even by Darwin. The two differed only on the relative significance of selection and use-inheritance, Darwin thinking selection was the more powerful, Spencer subordinating selection to the Lamarckian effect. How was it possible, then, for Lamarckism to emerge toward the end of the century as the basis for an explicitly anti-Darwinian view of evolution? To understand how this could happen, we have to realize that the same basic idea can be presented ("spun" is the modern term) in very different ways. What the later generation of neo-Lamarckians rejected was not just natural selection—it was the whole package of evolutionism associated with Darwin and Spencer. Spencer's own version of use-inheritance was seen as essentially materialistic, and hence as not all that different from natural selection. The whole process was driven by individual selfishness, a desperate drive to succeed in the struggle for existence. If animals responded to their environment by developing new characters, this was a mechanical reaction driven by unconscious processes in the body. For those who disliked the assumption that conflict-driven individualism was the sole mechanism of progress, Spencer's version of Lamarckism was as distasteful as the selection of random variation.

The anti-Darwinian version of Lamarckism emerged as part of a wave of opposition to the whole materialist program with which Darwin, Spencer, and Huxley were associated (Turner, 1974). Huxley had tried to promote the name "scientific naturalism" for this program, but for most of his contemporaries it was little better than materialism. The opponents preferred to believe that the universe was driven by processes which reflected an underlying moral purpose. Many of them were openly vitalist in their view of life, insisting that living bodies were animated by a life force that could produce purposeful effects beyond the capacity of any material structure. To these thinkers, the Lamarckian effect was a sign that living things could control not only their own destinies, but also the future evolution of their species. A theory that was in some ways very similar to Spencer's thus took on very different overtones. One way of trying to map this difference onto science was to argue that the directing power of the life force would impose a more orderly pattern of development on evolution than anything driven by so haphazard a process as natural selection.

There was no implication here that evolutionism itself should be rejected. This period saw an immense level of interest in efforts to reconstruct the history of life from fossil and other kinds of evidence (Bowler, 1996). Many of the scientists involved were themselves determined to find an alternative to materialism—there was no sense of a confrontation between science and forces opposed to the rational study of nature. The opponents of materialism rejected the Darwinian-Spencerian model of evolution, but they were only too happy to build the basic idea of progressive evolution into their own ideology. For them, progress meant the ability of living things to transcend the demands of the purely material world—it was not a mere by-product of economic activity.

The opposition to materialism came as much from a philosophical and moral perspective as it did from formal religion. Some of the anti-materialists were indeed deeply religious people. But others opposed materialism because they found it morally distasteful, even though they had little time for conventional religion. Even T. H. Huxley eventually became disenchanted with the social Darwinism implicit in the Spencerian approach. In his lecture "Evolution and Ethics" of 1893 (in Huxley, 1894; see Helfand, 1977; Paradis, 1978) he accepted that human morality had nothing to do with the struggle for existence which had driven natural evolution. The neo-Lamarckians simply took this argument a step further and queried whether or not evolution was, in fact, driven by struggle. Lamarckism didn't have to involve struggle (whatever Spencer might think) and thus offered the possibility of constructing an evolutionary worldview in which values such as altruism were built into nature. This rejection of struggle is especially visible in the writings of literary figures such as Samuel Butler and George Bernard Shaw, but many biologists also hoped to liberate science from the materialist paradigm. There was an uneasy alliance between these philosophical anti-materialists and the more liberal Christian theologians who were trying to "modernize" religion to make it more compatible with contemporary values.

There were scientific grounds on which the anti-materialist position could be based, including the latest theories in physics. One of the founders of the new science of energy, thermodynamics, was William Thomson, later Lord Kelvin (Smith and Wise, 1989). He had been raised a Presbyterian, and although he later adopted a more liberal religious position, he remained committed to the belief that the universe was a divine creation powered by an initial store of energy implanted by God. Although not opposed to the basic idea of evolution, he found it difficult to accept the Darwinian view that it was driven by purposeless forces. As early as the 1860s Thomson began to use his physics to attack a crucial foundation of Darwin's theory, his reliance on Lyell's estimate of a vast amount of time for the history of the earth. There would be no return to the old idea of a 4004 B.C. creation, but Thomson applied his science of energy flow to the earth itself to show that it could only be at most a hundred million years since the planet was a mass of molten rock (Burchfield, 1975). The interior of the earth is hot, and if there was nothing to maintain the internal heat, the whole planet must cool down at a rate the physicist can determine. Thomson's attack on Lyell's vast timescale was a coded attempt to undermine the credibility of Darwin's theory. He knew that Darwin needed vast amounts of time because natural selection must be an extremely slow process. Without thousands of millions of years in which to progress, evolution could not have produced the advanced species we now observe. If Lyell was wrong on time, then Darwin was wrong on evolution—it would have to be driven by something more progressive than natural selection.

Given Thomson's reputation as a physicist, this was a serious blow to the credibility of the selection theory. Many of the biologists who promoted alternative mechanisms cited his work as a reason for supposing that a more positive "drive" would be needed. The situation became even more critical as Thomson reduced his estimates of the earth's age even further toward the end of the century. Many geologists had accepted the original estimate of a hundred million years, but they protested when it was reduced even further. Only in the early twentieth century did it become apparent that Thomson's whole approach was flawed. The discovery of radioactivity revealed that there were processes that could produce heat deep in the earth over vast periods of time. The planet is not cooling down because the decay of radioactive elements such as uranium generates heat to replace what is lost into space. Soon geophysicists such as Arthur Holmes were using the decay of radioactive elements to provide a new estimate of the earth's age (Lewis, 2000). The answer came out very quickly to a figure remarkably close to what geologists still accept today—about four and a half billion (thousand million) years. Twentieth-century Darwinists would not have to worry about time. By the same token, an important if indirect argument in favor of the anti-Darwinian theories evaporated.

In the closing decades of the previous century, though, Kelvin's arguments seemed unassailable, and the search was on for a more purposeful mechanism of evolution. One of the most popular alternatives was the inheritance of acquired characteristics. Known as "neo-Lamarckism," this process became the basis for an explicitly anti-Darwinian view of evolution. It was also presented as an alternative to the Spencerian version of Lamarckism that was linked with Darwinism. Spencer agreed with Darwin that the crucial factor shaping the evolution of a species was its response to the external environment. The neo-Lamarckians approached the theory from a different direction which limited the role of the environment, and hence the effectiveness of the struggle for existence. One of their most popular arguments was based on the so-called "recapitulation theory" in which the development of the individual embryo passes through phases corresponding to the past evolution of the species (Gould, 1977). Although he realized that embryos preserve valuable clues about evolutionary relationships, Darwin had never been impressed with the claim that the adults of the lower animals correspond to immature phases in the development of the human embryo. An early human embryo exhibits some characters suggesting a distant relationship to the fish (because they were the earliest vertebrates), but it never looks like an adult fish. The neo-Lamarckians took recapitulation far more seriously, because they thought that evolution proceeded by *adding on* stages to the process of individual development. Characters acquired by the individual's efforts are inherited by being added on at the end of embryological development. Evolution becomes a process of linear addition, and the development of the embryo becomes the model for a progressive and goal-directed form of evolution.

At this point, the puzzled reader will no doubt ask: but surely Lamarckism still required adaptation to the environment—so why should its results be any less haphazard than those of natural selection? This was how Spencer had interpreted the theory, and for him progress was achieved by adding together a mass of small-scale changes that only generated an advance when averaged out over vast periods of time. The neo-Lamarckians took a different view: they focused on the initial step when members of the species first "discovered" a new habit that would generate useful characters. Once that habit is established, it shapes the whole course of evolution in a purposeful direction. The proto-giraffes who discovered that they could obtain food from the trees effectively marked out the whole future course of their species' evolution. Later generations continued stretching their necks upward, and their efforts added up to give the giraffe of today. Evolution consists of occasional episodes of innovation, followed by long periods in which it advances as though toward a predetermined goal. The whole approach seems much less harsh and less materialistic than the model proposed by Darwin and Spencer. Evolution is directed toward goals decided by the animals' creative choices.

Some neo-Lamarckians were so fascinated by their idea of pre-determined trends that they thought evolution could acquire a momentum that would drive the species too far along the chosen path. The gigantic horns of the so-called "Irish elk" were thought to have become so big that the animals couldn't support them and the species went extinct. This notion of nonadaptive "orthogenesis" highlights the difference between neo-Lamarckism and the Darwinian-Spencerian approach. To Darwin it was unthinkable that evolution could drive a species to extinction by promoting a character that was becoming maladaptive. The pressure of the struggle for existence would block such a tendency as soon as it appeared—any individuals affected would simply be eliminated. It was this willingness of biologists to consider mechanisms of nonadaptive evolution that led Julian Huxley to describe this episode as an "eclipse of Darwinism"—it was far more than an rejection of the selection theory, it was a rejection of the whole idea that evolution is the summing up of local changes policed by environmental pressure.

It was in America that the neo-Lamarckian movement became most clearly defined (Bowler, 1983, chap. 6; Pfeifer, 1965). Many American biologists, including Asa Gray and James Dwight Dana, accepted evolution but were disturbed by the capacity of the selection theory to undermine the argument from design (see Livingstone, 1987; Numbers, 1998). A group of naturalists and paleontologists influenced by the Harvard naturalist Louis Agassiz openly rejected the Darwinian theory in favor of Lamarckism. Agassiz himself remained committed to the idea of divine creation (Lurie, 1960), but his younger disciples realized that science needed the basic idea of evolution to make sense of the development of life on earth. The paleontologists Edward Drinker Cope and Alpheus Hyatt, and the entomologist Alpheus Packard, all sought to develop a vision of evolution which would retain Agassiz's sense that nature exhibits a coherent pattern, more regular than anything permitted by natural selection. Cope and Hyatt were both active in using the fossil record to reveal evidence of linear trends in evolution, which they could claim as evidence in favor of neo-Lamarckism and orthogenesis. Cope is remembered as the eccentric fossil-hunter who engaged in a bitter rivalry with Othniel C. Marsh over the exploitation of the rich fossil beds of the American West (Wallace, 1999). Their fossil-hunting teams fought, sometimes literally, over the most productive beds in territory that was only just being seized from the Native Americans. In 1890 their hostility exploded onto the pages of the *New York Herald,* each side accusing the other of dishonesty and slipshod science.

Cope certainly described many new dinosaurs and other extinct species, although often so hastily that he made mistakes. He also became known for his efforts to show that evolution was governed by linear trends that could not be explained in Darwinian terms. Often, he claimed, one could see evidence of several related groups evolving in parallel, as though driven toward the same goal by some internal force. Parallelism, like the claim that some trends ended up with overdeveloped characters and extinction, was something else that the Darwinians repudiated—what could possibly drive unconnected species living in different locations in the same direction? But Cope thought he could detect such linear trends in the evolution of many groups, including the fossil horses with which Marsh had so impressed T. H. Huxley. Hyatt saw similar trends in fossil ammonites and other cephalopods, which often ended in the production of bizarre characters as a prelude to extinction.

Cope was explicit about the moral and religious agenda behind American neo-Lamarckism. He had been brought up as a Quaker, and remained concerned with theological issues throughout his life. Although he did not study under Agassiz, his first ideas on evolutionism were very much intended to synthesize Agassiz's vision of creation with the new evolutionism. For Agassiz, the parallel between the development of the human embryo and the progressive sequence in the history of life on earth was a sign that both were governed by the same underlying pattern, a pattern emanating from the mind of God. Cope merely argued that the divinely implanted pattern might unfold in a nonmiraculous way in both cases. In effect, he repeated the argument used in Chambers's *Vestiges* to suggest that the laws governing reproduction can impose a program of development onto the evolution of the species. In his first paper of 1867 he did not even believe that the sequence was determined by adaptive pressures; the path of evolution was "conceived by the Creator according to a plan of His own, according to His pleasure" (Cope, 1868: 269; his papers are collected in Cope, 1887a). Later he accepted the Lamarckian view that it was newly adopted habits which established the trend worked out in the later phases of the group's evolution. In a book entitled *The Theology of Evolution* (1887b) he explored the religious implications of this position in detail. Cope was a vitalist—he believed that living things were governed by nonphysical forces, especially a growth-force he called "bathmism." It was this which gave living organisms the ability to make a creative response to their environment and thus to determine their species' future evolution. In effect, the vital forces represented God's creative power delegated to the natural world. Like Kingsley's Mother Cary, God didn't need to make living things directly because He had given them the power to make themselves.

Cope's viewpoint was extended by another scientist, Joseph LeConte, who wrote more extensively on the religious and social implications of neo-Lamarckism (Stephens, 1982). His *Evolution and its Relations to Religious Thought* of 1888 promoted Lamarckism as the main mechanism of evolution in its early stages, but conceded a role for selection later on. LeConte thought that human consciousness was on a higher level than that of the animals, and that this had once again allowed Lamarckism full rein, since now we could more easily pass the benefits of our experience on to future generations. He was particularly anxious to show that the Lamarckian mechanism was best applied not by allowing nature to take its course (Spencer's free-enterprise model) but through the government actively promoting social policies that would work for the benefit of all. This anti-Spencerian social message was also developed by the sociologist Lester Frank Ward (Scott, 1976, and more generally on the debates sparked by social Darwinism, Hofstadter, 1959).

Lamarckism came increasingly under fire from genetics in the early twentieth century, but the American school's approach continued to be influential among paleontologists. Cope's disciple Henry Fairfield Osborn became one of the most powerful American biologists of the early twentieth century through his position at the American Museum of Natural History. Although conceding that Lamarckism itself was of dubious value, he continued to stress the model of evolution in which many parallel lines advanced together in the same direction (Regal, 2002). Osborn developed the idea of occasional bouts of "adaptive radiation"—a concept still in use today to denote the periods of rapid divergent evolution that seem to follow a mass extinction. At this point, evolution was truly creative, although Osborn thought that once the various groups were established, their further evolution was governed by rigid trends. Osborn was one of the foremost figures leading the scientific defense of evolution against the upsurge of creationism in the 1920s. He was particularly anxious to use his anti-Darwinian model to distance the theory from the image of unrelenting struggle promoted by the Spencerians.

Neo-Lamarckism also flourished in Europe, although less so in Britain than in some other countries. Not that the moral implications of anti-Darwinian evolutionism were ignored—in fact they were presented all too vigorously by the novelist Samuel Butler (Pauly, 1982; Willey, 1960). Best known for his novel *Erewhon*—an imaginative attack on the dangers of the machine-age culture—Butler at first considered himself to be a Darwinian, but after reading Mivart's *Genesis of Species* he saw the dangerous implications of the materialism inherent in the selection theory. But where Mivart saw design imposed on evolution by a divinely implanted trend, Butler followed the same logic as Cope by realizing that the Lamarckian theory allowed the creativity of the animals themselves to shape the course of evolution along beneficial lines. Unfortunately, he promoted this view in a series of books that were bitterly critical of Darwin himself, beginning with his *Evolution Old and New* of 1879. According to Butler, Darwin had not only pioneered a dangerous model of evolution—he had also concealed the fact that a better approach had already been proposed by Buffon, Lamarck, and his own grandfather Erasmus Darwin. By now, Darwin was a hero to the British scientific community, and this attack led to Butler being ostracized. Even so, he continued to attack the moral implications of the selection theory: "To state this theory is to arouse instinctive loathing; it is my fortunate task to maintain that such a nightmare of waste and death is as baseless as it is repulsive." (Butler, 1908: 308)

An important shift of emphasis took place in the early twentieth century under the influence of the French philosopher Henri Bergson. His *Creative Evolution* (translated 1911) became a rallying point for those opposed to old-fashioned materialism both in science and in religion. Bergson was explicitly a vitalist: he believed that the life force, the *élan vital,* progressed by struggling to overcome the limitations of brute matter. Evolution had no preordained goal, because it was impossible to predict the various ways in which living things would triumph over these limitations. Two very different outcomes were the insects, whose success was based on instinct, and the vertebrates, which progressed through individuals learning how to deal with obstacles. Humans were the highest product (so far, at least) of the latter trend. There were many scientists who found Bergson's ideas inspirational. The Scottish biologist J. Arthur Thomson, one of the most prolific writers on popular science in the early twentieth century, responded positively to his ideas and promoted a vision of living things rising above material needs (Bowler, 2005a). By this time it was increasingly clear that the Lamarckian mechanism was under threat from the new science of genetics. Thomson and his allies were not openly Lamarckians, but they had inherited the anti-materialist vision of the earlier generation. The emphasis was now on the creativity of life, not on the rigidly predetermined trends of orthogenesis.

A prominent exponent of creative evolution was the playwright George Bernard Shaw, who declared in the preface to his *Back to Methuselah:* "If it could be proved that the whole universe had been produced by [natural] selection, only fools and rascals could bear to live" (Shaw 1921: liv). He hailed Samuel Butler as a hero who had been dismissed by a scientific community obsessed with materialism. Shaw illustrates a growing tendency to present the late nineteenth century as a period completely dominated by materialistic Darwinism—a misconception too often repeated even today (Bowler, 2005b). In fact, as we have seen, Darwin, Huxley, and Spencer did not have it all their own way, even in science. There was a vast upsurge of anti-Darwinian feeling throughout the later part of the century, and the exponents of creative evolution were doing little more than repeat the arguments of the previous generation. Their moral objections to the selection theory were clearly expressed, and for a while it looked as though the vitalistic form of Lamarckism might serve as the basis for a scientific alternative to Darwinism. But by the time Shaw entered the debate, the inheritance of acquired characters was coming under fire from a new development in biology: Mendelian genetics.

GENETICS AND THE REVIVAL OF DARWINISM

There were a few supporters of the selection theory during the eclipse of Darwinism, and their work clarified issues central to the development of genetics. In the end, genetics would destroy the credibility of the main alternatives to Darwinism and provide a new foundation for the selection theory. In the short term, however, it was seen as yet another anti-Darwinian mechanism, and the process by which the two approaches were reconciled was quite convoluted. Traditionally we associate the origins of genetics with the breeding experiments conducted in the 1860s by Gregor Mendel, which were ignored at first but were "rediscovered" in 1900. There is some doubt as to whether or not Mendel himself appreciated the full implications of his techniques for the theory of heredity. But leaving these historical reassessments aside, we need to understand how a group of biologists began to rethink the mechanism of heredity and belatedly recognize Mendel's techniques for studying the transmission of characters as the units we now know as genes (for surveys see Bowler, 1989; Gayon, 1998; Olby, 1985; Provine, 1971). One of their chief inspirations was the anti-Darwinian theory of evolution by sudden jumps or saltations.

The most influential Darwinist of the late nineteenth century was the German biologist August Weismann. He was responsible for a conceptual innovation about the nature of heredity that undermines Lamarckism and leaves natural selection as the most plausible mechanism of evolution. Weismann anticipated the modern idea that the process of heredity works through the transmission of information from parent to offspring, and that this information is encoded in a material substance contained in the chromosomes of the cell nucleus. He called this substance the "germ plasm"—today we know it as DNA. Weismann's most controversial claim was that the transmission of information works in one direction only. The germ plasm passed on from the parents to the fertilized ovum contains all the information needed to develop the embryo of the new organism. But once the new organism is formed, there is no way in which subsequent changes in its body can be impressed on the germ plasm it carries for transmission to its own offspring. The inheritance of acquired characters is thus impossible: the weightlifter can develop his own muscles beyond the normal level, but the information corresponding to that modification cannot be impressed on his germ plasm. In modern terminology, the genetic code stored by the DNA will program the development of the new organism, but there is no mechanism by which changes to the adult can be encoded into the DNA in its reproductive cells.

Weismann declared that the Lamarckian theory was invalid and performed a famous experiment to show that when rats had their tails cut off, there was no tendency for the tail to diminish in future generations. Virtually everyone else at the time, Darwin included, thought that by docking the tail you would eliminate whatever produced the tail in the offspring, and some effect of the mutilation should be apparent in the next generation. Weismann held that the information for producing the tail was not generated in the parents' tails, but was encoded in their reproductive cells. Here it was isolated from all outside influences, even from the body that carried it. The Lamarckians responded by pointing out that their theory postulated the transmission of positive adaptations, not mutilations, but they found it hard to provide convincing evidence of the effect in the laboratory.

If the Lamarckian effect did not work, what was the mechanism of evolution? Weismann believed that his theory of heredity confirmed that Darwin had been right. Evolution could only proceed from the natural selection of random variations introduced into the germ plasm. Such variations might result from the simple recombination of existing characters by sexual reproduction. But in the long run there would have to be some process by which existing

characters were supplemented by new ones. There must be disturbances of the transmission process leading to modifications of the
germ plasm and hence of the characters it coded for. Since such disturbances could not be controlled by the body within which the
germ plasm was hosted, the results would be random, just as Darwin had assumed. Natural selection could then winnow out the
harmful characters and spread the occasionally useful ones into the
whole population.

Weismann pioneered an idea that would be embedded in modern genetics, the separation of the material responsible for the
transmission of characters (the germ plasm) from the characters
themselves. But in one respect he did not anticipate the model
employed by Mendel in his classic experiments. As a good Darwinian, Weismann thought of variation in terms of minute differences
spread through the whole population—which is, of course, exactly
what we observe in most species. Think of the continuous range of
variation in height among human beings, for instance. Among the
relatively small number of active Darwinians in the late nineteenth
century, the statistician Karl Pearson and the biologist W. F. R.
Weldon performed detailed studies of such continuous ranges of
variation in species as diverse as crabs and snails, and were even
able to show the effects of natural selection acting on such variation, although on a very small scale.

The one thing that these Darwinians did not anticipate—because it was alien to their theory of gradual evolution—was that
it might be possible to treat some characters as discrete units inherited on an all-or-nothing basis. This was the model used by the
Moravian monk Gregor Mendel in his classic breeding experiments
published in 1865. Working in his monastery garden, Mendel
crossed artificially bred varieties of the garden pea plant and was
able to show that here (unlike the situation in many natural species) there were distinct character-differences which did not blend
together in the hybrid offspring. The artificial varieties consisted

solely of tall and short plants, for instance, with none of a medium height (as there might be in a natural population). And when tall and short parents were crossed, the offspring were still not of medium height—they were all tall. When crossed again, the short character reappeared in a quarter of the second hybrid generation, giving Mendel's famous 3:1 ratio. Modern historians are not sure that Mendel intended his model to serve as the basis for a new science of heredity (see Bowler, 1989; Olby, 1985). His raw material, the artificial strains of peas, exhibited rigid differences quite unlike the variation seen in most natural populations. It is also doubtful that Mendel thought of material units (what we now call the genes) corresponding to the discrete characters. His experiments were so far out of touch with contemporary thinking on heredity that they went largely unnoticed until 1900, when the effects he had studied were "rediscovered" by two biologists, Hugo De Vries and Carl Correns, and hailed as the basis for a new theory of heredity.

What had changed to make the model based on unit characters more plausible? The answer to this question lies in the emergence of another anti-Darwinian theory, the idea that evolution proceeds not by gradual change, but in discrete jumps or "saltations." Even T. H. Huxley—never a very enthusiastic supporter of natural selection—had taken this possibility seriously. In the 1890s a number of biologists began to look for evidence of characters that could only have been generated as discrete units. The English biologist William Bateson published a strongly anti-Darwinian book, his *Materials for the Study of Variation* of 1894, which pointed out examples that the Darwinian theory could not explain. If, for instance, the flower of a plant species existed in a four and a five-petaled version, it was difficult to believe that the extra petal had been formed gradually over many generations. More likely there had been a sudden switch in the process of individual development, invoking an extra manifestation of the "instructions" for making the petal. Bateson's rejection of Darwinism reflected a general dissatisfaction with the theory among biologists who were becoming increasingly focused on laboratory work rather than field studies. Having worked on the evolutionary origins of the vertebrates, Bateson had decided that the most important questions posed by evolution theory could not be answered with the techniques available. His move to the study of variation and heredity was a deliberate reaction against the sometimes speculative attempts that had been made to reconstruct the evolution of life on earth.

The Dutch botanist Hugo De Vries thought that he had actually observed the production of discrete new characters in the evening primrose—he called them "mutations." De Vries's mutations were large-scale changes which generated a new species instantaneously, exactly the phenomenon anticipated by advocates of the saltation theory. His work was taken up enthusiastically, especially in America. One of its leading supporters was Thomas Hunt Morgan, who—like Bateson—had reacted against the speculative efforts to reconstruct ancestries and now used the saltationist model as a stick with which to beat the Darwinians. De Vries did at least admit that there would be a selection process operating among the mutated forms, but Bateson and Morgan denied that there was any role for adaptation in evolution. New characters were produced by mutation and simply established themselves as species, whether or not their characters were adaptive.

Bateson, De Vries, and Morgan all played significant roles in the development of genetics. The rediscovery of Mendel's results was made possible because biologists who had become convinced that characters were produced by discrete jumps were naturally inclined to expect that they would also be inherited as discrete units. Genetics—Bateson's name for the new science—emerged from the saltationist theory of evolution. It was thus presented as yet another nail in the coffin of Darwinism, undermining the whole foundation of the model based on continuous variations used by Darwin, Weismann, Pearson, and Weldon. The Darwinians in turn rejected genetics (Gayon, 1998; Provine, 1971). Each side dismissed the phenomena studied by the other as trivial. The geneticists thought that continuous variations were environmentally induced and not inherited—so they could play no role in evolution, which depended on the production of new, hereditable characters by mutation. The Darwinians thought the discrete characters studied by the geneticists were artificial products of human breeding.

The one point of agreement between the two schools was that both accepted Weismann's insistence that the characters relevant for evolution were produced within the germ plasm, or the genes, and that (whether small or large) they were unaffected by modifications of the adult body. Acquired characters could not be inherited. This position was taken up even by those geneticists who were reluctant to accept the idea that the gene was a material entity located on the chromosome. Bateson, who himself did not accept the materialistic theory of the gene, led the assault on the few biologists who were still trying to find experimental evidence for the Lamarckian effect. The theory upon which so many scientists, moralists, and religious thinkers had based their rejection of Darwinian materialism was now in serious trouble. If anything, the combination of genetics and the mutation theory was even harder to reconcile with any hope of seeing purpose in evolution. Although it made no use of the struggle for existence, it presented the origin of new characters as the result of nothing more than disturbances arising in the material structure of the gene. In modern terminology, mutations were just copying errors, and for the early geneticists this alone was responsible for evolution.

Within a couple of decades, though, the geneticists' rejection of natural selection had to be reconsidered. T. H. Morgan's studies of mutations and their inheritance in the fruit fly, *Drosophila*, showed that genuine mutations do not create new species; they merely generate new characters within the existing breeding population. And many mutations are quite trivial in their effect, not the huge leaps imagined by the saltationists. It was also becoming clear that the simple cases studied by the early geneticists, where a single genetic difference corresponded to two discrete states for a character, were not typical. Most characters, like height in the human population, are affected by a number of different genes, whose effects are constantly stirred up by recombination through sexual reproduction. Even Morgan at last began to admit that mutations were merely supplying the random variation that Darwin, Pearson, and Weldon had studied in wild species. In which case any new mutation producing a beneficial character would increase its frequency in the population, because those organisms carrying it would reproduce more vigorously. Those coding for a useless or harmful character would be kept down to a very low frequency. The idea of natural selection was beginning to reappear.

The basis for modern Darwinism emerged when a younger generation of biologists began to work out ways of studying the genetics of whole populations (Provine, 1971). This required statistical techniques of exactly the kind pioneered by Pearson: mathematical formulas that could represent the changing frequency of a vast number of genes circulating in large populations. Pearson himself remained reluctant to give genetics a central role in evolution theory, but his followers increasingly recognized that the population is a reservoir of genetic variation, on which natural selection acts by modifying the proportion of the genes. One advantage of this approach was that it undermined an objection that had been advanced by Mivart and others against the selection theory: if an evolutionary change requires the smooth integration of a number of different characters, how can they all appear at the same time if each is the product of a single random event? Population genetics showed that initially useless or even harmful genes produced by mutation can be preserved at a very low frequency in the population. If a particular combination eventually turns out to be useful in a new environment, the necessary characters don't have to be generated all together.

Population genetics was developed during the 1920s and 1930s by R. A. Fisher and J. B. S. Haldane in Britain and by Sewall Wright in America. All became convinced that natural selection was now the only viable mechanism of evolution. Fisher's book *The Genetical Theory of Natural Selection* (1930), and Haldane's more popular *The Causes of Evolution* (1932) helped to establish the basis for the new Darwinism. There were still further developments to be made, and it was not until the 1940s that the whole "modern synthesis" of Darwinism and genetics was in place (the term was popularized by Julian Huxley's book, *Evolution: The Modern Synthesis* of 1942; see Mayr and Provine, 1980). Huxley and the next generation of Darwinists adapted the model developed by Fisher to the field naturalists' insights about how geographical barriers isolate populations that may then move far enough apart to be established as distinct species. These developments take us into the later period dealt with in Chapter 5, but it is clear that by 1930 Darwinism was beginning to emerge from its eclipse.

The moral and theological implications of the re-emergence of Darwinism would be profound. The hopes of liberal Christians that a reconciliation with evolutionism could be worked out on the basis of Lamarckism were dashed. The possibility of using the theory of natural selection to attack any notion that humans were the product of divine purpose was back in play. But were these implications apparent to those who were founding the new Darwinism? Haldane, to be sure, was an opponent of organized religion who became a Marxist in the 1930s. But Fisher was a liberal Anglican and Julian Huxley, while rejecting belief in a personal God, remained committed to a vision of evolution that saw humanity as the product of cosmic purpose (Bowler, 2001; Greene, 1990; Ruse, 1996; Swetlitz, 1995). Both Fisher and Huxley were enthusiastic followers of Bergson's creative evolution—but where Bergson thought that the creativity of the life force was incompatible with Darwinism, they saw natural selection itself as an opportunistic process ready to exploit any opening provided by the changing environment. It was thus not immediately apparent why the revival of Darwinism should upset the hopes of liberal Christians. In the short term, at least, the threats to that synthesis would come not from science but from changes taking place within the wider culture of Western society.

UP FROM THE APE

There were also major developments in the study of that most controversial area of evolutionism, human origins. When Darwin published his *Descent of Man* in 1871 there were virtually no human fossils available to fill the gap between the hypothetical ape ancestor and modern humans. The 1890s saw the discovery of the first really important missing link, the so-called Java man or *Pithecanthropus erectus*, now known as *Homo erectus* and widely accepted as the ancestor from which our own species, *Homo sapiens*, evolved. For a variety of reasons, many evolutionists were not at first happy with this apparent solution to one of their major problems. They postulated hypothetical lines of evolution that bypassed the few known fossils. In this very sensitive area, scientific theories were shaped by existing preconceptions—and when the evidence did not fit the preconceptions, it was marginalized (Bowler, 1986; Reader, 1981). When the fist example of *Australopithecus* was found in South Africa in 1924, it was largely ignored—although later discoveries have confirmed that the Australopithecines are the earliest members of the human family. Only in the 1940s did something like the modern view of human origins emerge.

Darwin had suggested a radical vision of human evolution that located it in Africa. He pictured the first step in the separation from the apes as an adaptive shift based on standing upright to walk out of the trees onto the open plains. He implied that the expansion of the human brain was not the driving force of our evolution. It was a by-product of our increasing ability to use our hands for the manufacture of tools. Most of his fellow evolutionists preferred to believe that humanity originated on the plains of central Asia, and their commitment to the idea of inherently progressive evolution led them to assume that it must be the expansion of the brain which led the way in separating the first members of the human family from their ape ancestors.

At the time the most important human fossil was the Neanderthal specimen, discovered in 1857. As noted in the previous chapter, there was some speculation that this might be a valid link between humans and apes because of its heavy brow ridges and sloping forehead. But even T. H. Huxley had noted that the Neanderthal brain was fully as large as a modern human's, hardly what one would expect in an intermediate. Even so, there were some efforts to depict the Neanderthals as a very late stage in the ascent from the apes. Some of the "lowest" forms of modern humans were depicted as almost living Neanderthals.

Expectation that humans had originated in Asia, not Africa, led the Dutch anatomist Eugene Dubois to the island of Java, in what is now Indonesia, to look for earlier human ancestors. In 1891–92 he uncovered a skull and thighbone, which he attributed to a new species *Pithecanthropus erectus* (erect ape-man)—the generic name being borrowed from the hypothetical missing link postulated by Ernst Haeckel. The skull had a capacity approximately halfway between that of apes and modern humans, as one might expect in the intermediate form. But the thighbone indicated that *Pithecanthropus* had walked fully upright. This was what Darwin had predicted, but it was incompatible with the more common view that the expansion of the brain had preceded the development of a fully upright posture. Haeckel and some of his followers tried to fit *Pithecanthropus* into a sequence leading from the ape to the Neanderthals and on to modern humans. But most authorities, Dubois included, eventually decided that this line of development was a side branch leading only to extinction. It ran parallel to the more important line leading directly toward modern humans via as yet unknown links. The expectation that the brain led the way in human evolution thus led a whole generation of paleo-anthropologists to dismiss what we now regard as a key piece of evidence for the evolution of humanity. Most contemporary paleontologists were obsessed by the idea of parallel lines of evolution marching in the same direction, so they were hardly likely to object when the same model was applied to the human family.

The same preconceptions also explain the temporary enthusiasm for "Piltdown man," eventually exposed as one of the most notorious cases of scientific fraud. Discovered at Piltdown in the south of England in 1912, the remains consisted of a rather modern-looking human skull apparently associated with an ape-like jaw. It was later realized that they were fakes, the jaw that of a modern ape carefully stained to match the other remains (Weiner, 1955). There is an extensive literature on the fraud, mostly written from a "whodunit" perspective, but scientists have always felt embarrassed that so many anatomists and paleontologists were at first taken in (Blinderman, 1986; Millar, 1972; F. Spencer, 1990). By identifying the preferred theoretical model of the period, we can see why it seemed so plausible—Piltdown provided evidence for the hypothetical parallel line of human evolution, distinct from the brutish Neanderthals. The expectation that the braincase would be the first part of our anatomy to achieve a fully human form seemed to have been confirmed.

This progressionist vision of human origins was exploited both by rationalists seeking to undermine religion and by liberal religious thinkers anxious to see evolution as the unfolding of a divine plan. One of the leading experts trying to reconstruct the Piltdown fossil was Arthur Keith, a noted agnostic who wrote for the Rationalist Press Association. In 1927 Keith gave an address at the meeting of the British Association for the Advancement of Science in which he highlighted the evolutionary model of human origins and endorsed the materialist view that the mind was a mere reflection of nervous activity in the brain. In response Oliver Lodge argued for a spiritual evolution paralleling progress at the biological level. The resulting debate generated a flurry of newspaper headlines (Bowler, 2001). Keith's rationalist position was also promoted by popular writers such as H. G. Wells, who saw science and the scientific management of society as the only way forward.

The Piltdown affair shows that evolutionists saw the origins of humanity as a story of inevitable progress, with several lines of development independently being driven toward the human form via slightly different routes. A striking manifestation of the enthusiasm for parallel evolution at the time is Henry Fairfield Osborn's theory that humans emerged not from the apes but from a much earlier mammalian form which had ranged over the plains of central Asia (Regal, 2002). According to Osborn, perhaps the leading American paleontologist of the day, the similarities that had led scientists to assume a link with the apes were independently evolved in the ape and the human lines. Significantly, this theory was highlighted in the 1920s when the fundamentalist assault on Darwinism was starting to gather strength. It allowed Osborn to sidestep one of the most common emotional reactions against evolutionism, disgust at the implication that we are related to the apes. There was a widespread feeling (no longer shared by many today) that the apes were brutal and nasty, hardly the sort of creatures one wanted as ancestors. Osborn was able to use his theory to argue that this concern was unnecessary—evolutionism didn't imply that we were descended from the apes.

This picture of human origins began to crumble in 1924 when the young anatomist Raymond Dart discovered a new hominid fossil in South Africa, which he named *Australopithecus africanus*. Dart concluded that here was evidence to confirm Darwin's original speculations about human origins. We had evolved in Africa, not Asia, because our ancestors were indeed related to the African apes. The teeth showed that *Australopithecus* belonged to the human not the ape line, and there was indirect evidence that the creature had already walked upright. Yet the brain was scarcely larger than that of an ape. Dart's find was widely dismissed at first, but by the later 1930s another South African, Robert Broom, began to unearth adult specimens of Australopithecines confirming that they were indeed the first members of the human family. Broom, an eccentric figure originally trained as a medical doctor, had already made a name for himself studying fossils that showed the transition from reptiles to mammals. Now he turned his hand to unearthing fossil hominids, with spectacular results. Darwin and Dart had been right after all: our ancestors separated from the apes by an adaptive shift linked to locomotion, and the expansion of the brain had come much later. *Homo erectus* (Dubois's *Pithecanthropus*) had been the next major step, when the brain had expanded enough to allow substantial toolmaking and had given these early humans the capacity to expand out of Africa and across the Old World.

The details of how humanity acquired its higher mental and moral faculties are complex and still widely debated. But the old assumption that evolution is somehow programmed to push the development of the brain and mental powers toward the human level is no longer accepted. We acquired our big brains as an unpredictable consequence of a separation of the ape and human lines based originally on a change in posture. This new, and in some senses less optimistic, vision of human origins emerged in the 1930s and 1940s along with, and almost certainly encouraged by, the replacement of the old theories of parallel evolution by the modern Darwinian synthesis. But as we now switch to consider the religious and moral developments of the early twentieth century, we need to bear in mind that the model of evolution then in play was not the modern one. The non-Darwinian theories inherited from the eclipse of Darwinism were still active, and most of the scientists who defended evolutionism against the fundamentalists of the 1920s were not Darwinians in the modern sense. They were trying to promote a vision of evolution that had been carefully tailored to minimize its most disturbing implications. Only the more radical of the younger biologists were aware that the situation was changing, and that a later generation would have to deal with a resurgent Darwinism which made no concessions to the hope that evolution was driven by a purposeful trend.

MODERNISM IN THE CHURCHES

What did the churches make of the eclipse of Darwinism? For liberal Christians the fact that science itself seemed to have turned its back on materialism offered renewed hope of reconciliation. The wave of enthusiasm for Herbert Spencer's philosophy had encouraged some religious thinkers, especially in America, to leap on the bandwagon of progressionism. The basic idea of evolution became widely accepted, and even Spencer's quasi-Darwinian explanation of how it worked appealed to some theologians who saw how this embodied aspects of the Protestant work ethic. A few, including Henry Ward Beecher, proclaimed openly that this would entail the rejection of the Calvinist approach to Original Sin. Christianity showed us how to improve ourselves—it did not require us to give up hope and throw ourselves on the mercy of God. Nor was Spencer's philosophy the only source of such a reinterpretation of the human situation. Many who opposed materialism nevertheless saw struggle at one level or another as a vital part of the process by which the individual or the nation sought to improve itself. For thinkers such as Ralph Waldo Emerson, struggle and conflict could be part of a spiritually uplifting process (Lopez, 1996; Porte and

Morris, 1999). By the early twentieth century, there were many liberal Christians who were prepared to move in this direction, linking a progressionist view of biological evolutionism with their hopes for the future spiritual development of humankind. But not all the liberals who accepted evolution were willing to endorse so open a rejection of the traditional Christian message. Bishop Barnes's gorilla sermons of the 1920s were designed to press home the point and force liberal Christians to face up to the full consequences of their acceptance of a more "scientific" worldview. Barnes hoped to move the churches into line with twentieth-century thought—but by this time the rise of fundamentalism in America was already showing that traditional evangelicalism was starting to fight back.

The liberal movement in religion was sometimes known as Modernism. (I use the capital "M" to make it clear that this is very different from the modernism that swept through early twentieth-century art and literature—indeed Modernism in the churches was a manifestation of exactly what the modernists in art rejected, i.e., the assumption that rational thought was steadily uncovering the true picture of the universe). In the Anglican Church in Britain there was a Modern Churchmen's Union with a periodical, *The Modern Churchman*, expounding liberal opinions. Barnes was widely perceived as a leader of this movement, although he preferred to call himself a liberal evangelical. In America too there were active liberal movements in many of the churches, with preachers such as Harry Emerson Fosdick promoting a "new" Christianity shorn of traditional ideas that, it was claimed, were merely relics of the worldview prevailing when the scriptures were written down. Although the viewpoint of the Modernists on either side of the Atlantic was pretty much the same, there seems to have been relatively little interaction, perhaps because the circumstances were so different. For British Modernists, the movement represented the best hope of turning back a growing tide of indifference to religion. In America, by contrast, the churches still flourished, and the Modernists found themselves fighting hard to resist the fundamentalists' calls for a return to traditional Christianity.

To many Modernists, the eclipse of Darwinism offered the hope of working out a synthesis with science that avoided some of the worrying aspects of the Spencerian philosophy. It was all very well to celebrate the importance of individual enterprise as the driving force of evolution, but by the turn of the century Spencer's critics were accusing him of setting up selfishness and ruthlessness as the basis on which nature was supposed to operate. There was talk now of a social Darwinism, which drew upon the Darwinian/Spencerian view of nature to argue that struggle and competition were inevitable and justified, whatever the consequences for the weak (Bannister, 1979; Hawkins, 1997; Hofstadter, 1959). The critics were wrong to see Spencer as rejoicing in the destruction of traditional morality, but it is easy to see how his worship of unrestrained free enterprise could be interpreted in this way. There was a growing sense that if evolutionism was to be acceptable as an explanation of how we got our moral values, something less committed to the "struggle for existence" would be preferable. The non-Darwinian theories of evolution offered just such an alternative. Neo-Lamarckism didn't need struggle and could be seen as a process designed to enhance the values of cooperation and self-sacrifice. Evolution was more than a compilation of individual acts of survival—it was an organized process in which trends drove onward toward a predictable goal. For the Modernists in religion, these nonmaterialistic versions of evolutionism were a sign that science was abandoning the more worrying aspects of the program set up by Huxley and Spencer.

One early indication of these changes was the immense success enjoyed by the writings of the Scots Presbyterian Henry Drummond (Moore, 1985b). Drummond began his career as an evangelical and gained an immense reputation for his ability to spread the Christian message. But he also came under the influence of Spencer, and turned his talents to promoting a synthesis that would preserve the

Christian message within the worldview of modern science. His *Natural Law in the Spiritual World* of 1883 argued that the rule of law governed human spiritual life just as it did the operations of nature. The human spirit must relate itself to God just as the living organism must adjust to its environment. Evolutionism was taken for granted—there was a continuous progress operating at both the biological and the spiritual levels. Drummond's hugely popular *Ascent of Man* of 1894 made it clear that the laws at work did not operate through ruthless competition. According to Drummond, nature was designed to promote the "struggle for the life of others." Evolution worked not by selfishness but by enhancing the cooperative instincts that can be found in almost all animal species, and which eventually gave rise to human morality. Altruism was built into the very process of evolution, so it was no surprise that the end product was a species with a highly developed moral sense such as our own. Like the neo-Lamarckians, Drummond took the highest elements of human character and built them into nature itself. This made it easier to see humans as a product of natural evolution without thereby degrading ourselves. It also made it easy to see evolution as the working out of a process instituted by a wise and benevolent God.

Drummond was no scientist, and he offered little evidence to back up his claim that evolution worked on the basis of altruism. How exactly did cooperation replace the struggle for existence as the mechanism of evolution? This question was answered in part by another popular writer, Peter Kropotkin, in a series of articles eventually collected into a book under the title *Mutual Aid* (1902). Kropotkin was a Russian prince who became an anarchist and was exiled to London. He claimed that before he was forced to leave his homeland, he had observed how animals exposed to the harsh climate cooperated in order to survive. It was only natural that evolution would enhance the cooperative instincts, not the selfish ones as Darwinism would lead one to expect. Significantly, Kropotkin appealed openly to Lamarckism to explain how the learned habit of cooperation could be transformed into an instinct for altruism.

Drummond's writings sold like hotcakes, and his efforts to promote a very liberal form of evangelical Christianity struck a chord with many younger believers anxious to feel that they could retain their faith without losing touch with the modern world. In effect, Drummond updated the message promoted already by preachers such as Beecher, but where Beecher left Spencerian individualism intact, Drummond translated the evolutionary message into the language of non-Darwinian evolutionism. This made it easier for his followers to feel that they were embracing a vision in which the higher moral values played an integral role. There was little room left for the old vision of a sinful humanity in need of redemption, but many young people felt that this was a religion that would still give meaning to their lives. Traditional evangelicals were appalled, but as yet their sense of betrayal was not focused on the particular issue of evolution.

The liberal evangelical message continued to excite enthusiasm through into the early decades of the twentieth century. A good illustration of this is the surge of excitement surrounding the "new theology" preached by the Congregationalist minister Reginald Campbell at the City Temple in London in the early years of the new century (Bowler, 2001). This was the most influential nonconformist pulpit in Britain, and for a time there was standing room only as people packed in to hear Campbell preach. His book *The New Theology* came out in 1907, in response to criticisms from more conservative members of his community. The main influence on Campbell was not Spencer but German idealist philosophy, which made it easier for his teaching to mesh with the vision of a creative life force promoted by Bergson. He stressed that his new theology was fully compatible with the latest views of the scientists who, he felt, were turning their backs on materialism. God was not a distant, transcendent Creator—He was immanent within the world, the driving force of a spiritual progress that had created humanity. More immediately, he was immanent within each of us, so that we could all play our part in the advance of humankind toward spiritual perfection. Sin was a failure to recognize one's place in the universal scheme, not a relic of the Fall requiring external salvation. Christ exemplified the future perfection of humanity. His divinity was merely an extended version of the divine spark in each of us.

Like Beecher's earlier rejection of the idea of Original Sin, Campbell's new theology brought home the extent to which acceptance of the idea of evolutionary progress undermined some of the most original tenets of Christianity. This can be illustrated through some of the contemporaries with whom Campbell identified himself. He praised the work of the noted physicist Oliver Lodge, who was an enthusiast for spiritualism and promoted a vision of souls evolving toward perfection in the next world, just as life had progressed in the course of evolution. Even more daringly, Campbell invited the playwright George Bernard Shaw to speak at the City Temple. Shaw was known as an advocate of creative evolution and a vitriolic opponent of Darwinian materialism—but he had also campaigned openly against Christianity. There was indeed a close parallel between the new theology and Shaw's vitalist evolutionism, but this drove home just how far Campbell had moved from the traditional Christian position. By the time he published his book, conservatives within his own church were expressing disapproval of his extreme liberalism. Campbell himself eventually had second thoughts and returned to a more orthodox form of Christianity within the Anglican church.

Whatever the hostility of conservatives, Modernism remained the most innovative movement in British theology through into the inter-war years (Clements, 1988; Stephenson, 1984; Bowler, 2001). Liberal theologians joined hands with scientists such as J. Arthur

Thomson, who were trying to preserve the antimaterialist view of nature inspired by Bergson's philosophy. There were other innovations too which seemed to offer hope. In 1923 the psychologist Conway Lloyd Morgan published his *Emergent Evolution,* promoting the idea that there were preordained stages in evolution at which entirely new properties appeared. Life itself had "emerged" from chemical evolution, whereas mind emerged with the higher animals and spirit with the appearance of the first true humans. Philosophers such as Samuel Alexander and Roy Wood Sellars endorsed the idea of emergence, encouraging liberal theologians to believe that key stages in progressive evolution were somehow preordained by the Creator. Alfred North Whitehead's *Process and Reality* appeared in 1929, offering a teleological vision of the world in which the emergence of human values played an integral role. Few could make sense of Whitehead at the time, but it was obvious that here was a fundamental challenge to the old materialism once associated with science. Liberal theologians sensed that it might offer a validation of the evolutionary perspective that was integral to their version of a transformed Christianity.

In Britain the Modern Churchmens' Union strove to promote a liberal view very similar to Campbell's new theology. The Modernists were convinced that the only way to save Christianity from the decline now affecting churches all over Europe was to make it compatible with those aspects of modern thought that the vast majority of people now took for granted. This meant taking on board the worldview of science, including the idea of evolution. It was from this background that Ernest William Barnes preached what the popular press called his "gorilla sermons" from the pulpit of Westminster Abbey in the early 1920s (reprinted in Barnes, 1927; see J. Barnes, 1979; Bowler, 1998, 2001). Unlike most clergymen, Barnes knew his science—he had taught mathematics at Cambridge before becoming ordained. He eventually published a massive survey, *Scientific Theory and Religion* (1933), which contained highly technical appraisals of the latest developments in physics. He was convinced that God governed the world solely through law—miracles were unacceptable in a scientific age and were not required for the foundations of Christianity. This meant rejecting traditional beliefs such as the virgin birth of Christ and the Resurrection. He campaigned openly against the Catholic belief that the consecrated wafers in the Eucharist or Mass acquire a spiritual essence. For Barnes, as for many evangelicals, the bread plays a purely symbolic role in the ritual. Like Campbell, Barnes seems to have regarded Christ not as the savior in the traditional sense, but as a great teacher who should be revered as a guide to what humanity could become in the future.

Evolution was a necessary part of this worldview, symbolized by Barnes's open call for us to recognize our ancestry in the apes. Humans had been produced by progressive evolution as part of the divine plan built into nature, and were now expected to take charge of their own future progress in order to reach the goal of spiritual perfection. What attracted the headlines was that Barnes drove home the need to recognize that this new approach to religion required the rejection of the old idea of Original Sin. The sermons he preached as Canon of Westminster, and later as Bishop of Birmingham, complained that Christians had tended to pay lip-service to the idea of evolution without fully confronting its implications for their faith. Everyone seemed to accept that evolution could be seen as the unfolding of a divine plan, and hence that humans still had a crucial role to play in God's creation. With a few exceptions, though, most theologians had refused to acknowledge that if we were the products of progressive evolution, and were expected to progress further in the future, then the Genesis story of a Fall from an original state of perfection had to be rejected. Human sinfulness was simply the relic of our animal ancestry that we now had to overcome, not the sign of a separation from God that could only be redeemed through the sacrifice of His son. Evolution confirmed

> that much that is evil in man's passions and appetites is due to natural instincts inherited from an animal ancestry. In fact, man is not a being who has fallen from an ideal state of innocence: he is an animal slowly gaining spiritual understanding and with the gain rising far above his distant ancestors. (Barnes, 1927: 312–313)

Here Barnes's evolutionism linked with his refusal to accept the miracles of the virgin birth and the Resurrection. To most traditionalists the whole Modernist package was a complete betrayal of the message that had sustained Christianity throughout its long history.

In his gorilla sermons Barnes was as vague as most contemporary theologians about the actual mechanism of evolution. Most liberal Christians simply assumed that some progressive, purposeful process was at work under the guidance of a creative spark which God had built into nature. In the early years of the twentieth century it was still possible to believe that the materialism of the Darwinian selection theory was not a threat, because science itself had discovered alternative mechanisms of evolution that were compatible with the idea that the process was driven by an underlying purpose. Another prominent Modernist in the Anglican Church, Charles Raven, campaigned openly in support of Lamarckism and against the new science of genetics, which was trying to undermine its credibility (Dilliston, 1975). In books such as his *Creator Spirit* of 1927 he called for a revival of natural theology based on the belief that the divine spirit was actively at work within evolution. Lamarckism was crucial because it allowed the individual to shape not only its own life, but also the future course of its species' evolution. Raven appealed to new developments in philosophy, including the thought of Whitehead, to free science from the legacy of old-fashioned materialism. As an enthusiastic naturalist and ornitholo-gist (he pioneered the use of photography to study bird behav-ior), he deplored the soulless laboratory work of the geneticists, which reduced animals to the level of mechanical puppets driven by their genes.

Unfortunately, it was the geneticists who were increasingly win-ning the battle within biology. As Barnes himself became aware, in their enthusiasm for the idea that evolution was driven by pur-poseful forces, the Modernists had backed the wrong horse. By 1930 Lamarckism was largely dead in science. If liberal Christians wanted to retain any credibility with the biologists, they were going to have to deal with the geneticists, and with the new generation of Darwinians who were reformulating the theory of natural selec-tion. It turned out that Barnes had actually taught one of the archi-tects of the new Darwinism, R. A. Fisher, when the latter had been a student at Cambridge. Fisher sent Barnes a copy of his *Genetical Theory of Natural Selection* of 1930, and Barnes tried to incorporate the new Darwinism into his 1933 survey of the relationship be-tween science and religion. Fisher was himself a liberal Anglican who seemed able to accept selection as a creative process of the kind that an (admittedly very remote) Creator might have insti-tuted. Barnes did his best to accommodate the possibility that evo-lution worked through the natural selection of essentially random mutations, but even he was not entirely comfortable with the pros-pect. Working out how to synthesize Christianity with the new Darwinism was going to take some time. In the meantime, Raven and many other liberal theologians continued to hope that the old non-Darwinian ideas could be salvaged.

Across the Atlantic, Modernism found itself with a different bat-tle to fight. Where the churches in Europe were steadily declining in influence, in America they continued to flourish. In the north-ern cities, which had become wealthy from industrial development, there was little sense that science was a threat to religion. Even in

Darwin's time, the ability of theologians such as Beecher to reconcile liberal Christianity with Spencer's social philosophy meant that evolutionism had become respectable. In the early twentieth century the Modernist movement continued to develop within the various Protestant denominations, at least in the big cities. Theologians such as Shailer Mathews and preachers such as Harry Emerson Fosdick insisted on the need to adapt the faith to the conditions of the modern world, including the scientific view of nature. Here, as in Britain, the anti-materialist trend in science, including ideas such as creative and emergent evolution, was seen as a boon to liberal Christians anxious to distance themselves from any hint that their approach might destroy faith in the existence of a Creator. But from the early years of the new century, the Modernists found themselves under threat from a resurgent traditionalism. The movement that became known as fundamentalism derived its support from those who had gained little from industrial progress and feared the erosion of family values that could all too easily be seen to flow from social Darwinism. Soon the theory of evolution would become one of the most visible arenas of conflict between the two versions of Christianity.

In America as in Britain, the liberal theologians formed an alliance with scientists anxious to show that their theories did not promote materialism. At the influential Princeton Seminary, Charles Hodge had rejected Darwinism as incompatible with design, but James McCosh and B. B. Warfield both promoted a reconciliation with a less materialistic view of evolution (see the modern editions of Hodge's *What is Darwinism* (1994) and Warfield's writings (2000); also Livingstone, 1987; Numbers, 1998, and more generally on evangelicals and science Livingstone, Hart, and Noll, 1999). The Modernist movement was still active in the 1920s. The historian Edward B. Davis (2005) has revealed a series of pamphlets on "Science and Religion" edited by Mathews in which he and Fosdick joined with a group of religious scientists to argue for an evolutionary view of nature that did not—as the fundamentalists alleged—endorse atheism. Issued by the American Institute of Sacred Literature (linked to the University of Chicago) and partially funded by the Rockefeller Foundation, the pamphlets were widely circulated in order to counter the threat of creationism. The biologist Edward Grant Conklin, the physicist Robert Millikan, and the geologist Kirtley Mather wrote pamphlets in the series, all promoting the view that the argument from design could still be applied within the evolutionary worldview of modern science.

Shailer Mathews, who edited this series, was perhaps the most active Modernist among the academic theologians. Mathews was a Baptist, and trained at Colby College and the University of Berlin. He began teaching New Testament history at the University of Chicago in 1905 and served as Dean of the Divinity School from 1908 to his retirement in 1933. In addition to pamphlets, he edited popular magazines and wrote several books urging the need for a reformulation of Christian belief, including *The Church and the Changing Order* (1907) and *The Faith of Modernism* (1924). The old-fashioned reliance on the supernatural was rejected, and evolutionism taken for granted, although Mathews acknowledged that the scientists were sometimes too dogmatic on the subject. The title of his autobiography, *New Faith for Old* (1936), encapsulated his determination to establish the foundations of a new form of Christianity purged of the ideas that made it unacceptable to the worldview of modern science. Mathews was one of the witnesses the defense hoped to call in the trial of John Thomas Scopes, and his views on the need to reinterpret scripture to allow room for the idea of evolution were widely quoted at the time.

Harry Emerson Fosdick was an immensely influential preacher and perhaps the best-known Modernist in the country. In the early 1920s he was fighting the fundamentalists from the pulpit of the First Presbyterian Church in New York. Fosdick argued against the whole idea of the Bible as a text that had been, in effect, dictated by

God and which hence had to be taken absolutely literally. It was not an inerrant text, but a product of the human experience of the divine, intended to focus our attention onto the value of the human personality in its relationship to God. Fosdick had been an evolutionist from the beginning—his family were deeply religious but had never seen any contradiction with the new ideas of science (Fosdick 1956: 49). Like many Modernists, though, he drew a distinction between evolutionism and Darwinism. Natural selection was only one theory of how the process might work, and many scientists accepted rival ideas that were far more compatible with design. Fosdick noted the writings of the Scottish biologist J. Arthur Thomson as a popular expression of this position. The idea of creative evolution effectively allowed us to see one aspect of God as immanent within nature, even if He also transcended the created universe. This reflected "the theistic evolutionists' view of an indwelling, purposeful Power, the Creative Spirit of the Living God unfolding, by slow graduation across measureless ages. . . ." Here was a cosmic evolution "slowly bringing forth life crowned with the possibilities of man" (Fosdick, 1926: 126). With his acute sense of the importance of the human personality, Fosdick appealed to the concept of emergence to justify the view that the soul was something more than a product of biological evolution. But he accepted the view, also promoted by Barnes in his gorilla sermons, that human sinfulness was a relic of our animal ancestry still capable of undermining our striving for something better.

Fosdick's career underwent a dramatic turn in 1924, symbolizing the rising threat posed by fundamentalism to the liberal cause. By this time the campaign that would lead to the Scopes trial was well under way, and Fosdick was actively preaching against the anti-evolutionists. But the Presbyterian General Assembly was captivated by the rhetoric of William Jennings Bryan and a movement to muzzle Fosdick built up through 1923 and 1924. In March, 1925 he left the

"Old First," vast crowds turning out to hear his final sermon, and moved to Park Avenue Baptist Church, where the influence of John D. Rockefeller ensured that a more liberal theology would prevail (Fosdick, 1956: chap. 7). In the final section of this chapter, we turn to consider the revival of traditional Christian values that had begun to turn the tide against liberalism.

THE RESURGENCE OF TRADITION

The emergence of what became known as fundamentalism in early twentieth-century America is the most obvious example of a broader change taking place within the Christian churches. Toward the end of the previous century, liberal views had commanded the attention of most articulate clerics, and this movement continued in the form of Modernism. But there had always been those whose views had kept more in line with traditional Christian beliefs and values, and now they were beginning to make their voices heard. The new century saw the Modernists' faith in the inevitability of progress checked by calamities such as World War I and the economic depression of the 1930s. It became less easy for liberal Christians (let alone secularists) to claim that history showed the continued ascent of humankind, following the pattern established throughout the history of life on earth. Traditionalists saw evolution as a symbol of the harmful effects that new ideas had had on morals and society. But the Protestant churches of America were not the only locus for a reassessment of tradition. The Catholic Church had turned against Modernism in the late nineteenth century, and it was only with the greatest of difficulty that Catholic theologians made room for a very limited form of evolutionism. Outside the ranks of the more evangelical Protestants, concerns about the collapse of moral values led to a resurgence of traditional views on the sinfulness of humanity even among intellectuals. In the 1930s the neo-orthodoxy of Karl Barth and Reinhold Niebuhr sidelined natural theology to emphasize the total separation of God from His fallen creation.

The key question for most late-nineteenth-century Christians, at least within the educated classes, had been that of design. Darwinism threatened the argument from design, and that is why so many people preferred non-Darwinian theories of evolution, which left more room for order and purpose in creation. The idea of progress allowed even the origin of the human soul from a lower order of creation to be seen as part of the cosmic purpose. The return to traditionalism saw the argument from design pushed into the background. What mattered to those who saw the idea of progress as a threat were the traditional family values that, they felt, were bring eroded by the new ideas, and that could only be salvaged by a return to belief in a God who had established those values and would hold us to account if we did not maintain them. Humans were something more than animals, and it was religion which ensured that we would not be tempted to model our behavior on them. For Protestants, at least, the Bible began to re-establish itself as the bulwark against atheism and anarchy. To begin with, there were still many evangelicals who saw no reason to treat every word of scripture as infallibly true, as long as the message proclaiming our need for divine guidance and salvation was upheld. But the need to defend the authority of the sacred text was beginning to reassert itself, paving the way for the emergence of a later generation to whom defending the accuracy of Genesis was paramount. All these moves made evolutionism an increasingly obvious target.

The Monkey Trial of John Thomas Scopes has come to symbolize the first phase of the attack on Darwinism, paving the way for the creationist movement of the later twentieth and early twenty-first centuries. But this episode has become so surrounded in myth that it has taken a great deal of work by modern historians to expose a more accurate picture of what happened in Dayton, Tennessee, in 1925. The trial was neither the start nor the climax of this first wave of opposition to the teaching of evolution in the schools, although it certainly got more publicity than anything else. The popular image of the fundamentalist William Jennings Bryan being demolished by the debating skills of the defense attorney, the agnostic Clarence Darrow, to the applause of all the big-city newspapers, is firmly entrenched in the popular imagination. Opposition to evolutionism was not defeated in the Monkey Trial and it continued to influence the American educational system until challenged again by the Darwinists of the 1950s and 1960s.

This reminds us of yet another twist in the story: for all that Bryan railed against the evils of social Darwinism, there were very few real Darwinists offering to support the defense of Scopes. In 1925 the eclipse of Darwinism defined the position of most working biologists, and the evolutionism on offer was not the materialist nightmare that Bryan conjured up—indeed the scientists were anxious to ensure that this image could not be sustained by their work. The eclipse of Darwinism in science ended only in the 1930s with the slow emergence of the selection theory as the dominant force in scientific biology. Thus the first phase of the Christian backlash against evolutionism coincided with the last phase in the attempt to create a non-Darwinian vision of evolution that would save the Modernists' faith in progress.

Before charting the rise of Protestant opposition to evolutionism, it is worth noting the more sustained holding action against the theory maintained by the Catholic Church. Catholic responses to the original Darwinian debate were complex but mostly negative, especially on the topic of the origin of the human soul (Appleby, 2001; Paul, 1974). As noted in the previous chapter, one of the most active anti-Darwinian biologists was the Catholic anatomist St. George Jackson Mivart. But Mivart was not opposed to evolution—indeed he was actively trying to convince the Church that it should take theistic evolution seriously. He remained firmly opposed to the idea that the human soul could have evolved from lower mentalities. He seems to have had some success in this project at first, but in the last decade of the nineteenth century the Church's attitude toward Modernism in general and the idea of evolution in particular hardened (Brundell, 2001; O'Leary, 2006). Mivart was excommunicated just before his death in 1900, although this was not explicitly for his evolutionism (Gruber, 1960).

In America, the Catholic priest John Zahm, who taught at the University of Notre Dame, published his *Evolution and Dogma* in 1896, endorsing a position very similar to Mivart's. The book was condemned by the Congregation of the Index, a Catholic committee in Rome charged with producing an index of books prohibited to Catholic readers, and was withdrawn. The condemnation was never actually published, however. This set the pattern for the Church's attitude during the early decades of the new century: there was no public rejection of evolutionism, but every effort was made to check the spread of the idea even at the purely biological level. In Britain, where the Protestant anti-evolutionary movement was never very strong, popular Catholic writers such as Hilaire Belloc and G. K. Chesterton were among the most visible opponents of rationalists such as H. G. Wells and Arthur Keith (Bowler, 2001).

The Church's unofficial line against evolutionism was challenged in a book by the Belgian geologist and priest Henri de Dorlodot (1925; see De Bont, 2005). Dorlodot endorsed the evidence for biological evolution and argued (as had Mivart) that there was nothing in the writings of the early Church fathers to suggest that they took the Genesis story of creation literally. Creation could take place through a divinely guided evolutionary process. It would therefore be acceptable for a Catholic to believe that evolution had formed the body of Adam, the first man, even if the soul was divinely implanted at that stage. Dorlodot promised another book on the question of human origins. This was never published, because the

Church tried to silence him and to force him to retract his arguments. He was never formally censured, however, and over the next decade it became apparent that liberal Catholics would be allowed to believe that the human body was a product of natural evolution, provided they continued to accept the supernatural origin of the soul. A book outlining this position was published by Ernest Messenger under the title *Evolution and Theology* in 1931. It may be noted, though, that such a position was not formally acknowledged until 1950, and only in recent decades has the Church openly supported even this limited form of evolutionism. The claim that the human spirit could have evolved along with the body was (and still is) unacceptable, as can be seen from the muzzling of the Jesuit priest and paleontologist Pierre Teilhard de Chardin. As we shall see in the next chapter, Teilhard's teleological cosmic evolutionism attracted widespread attention when his books were posthumously published in the 1950s, but in the prewar years he was not allowed to speak out openly.

The Catholic Church thus took a strong line against what had always been the most disturbing aspect of evolution, the implication that humans are merely improved animals. But concern for the literal meaning of Genesis was never an important aspect of its position. The Church has always regarded itself as having the authority to interpret scripture in the light of tradition, and its tradition has never included literalism. The fundamentalists who became active within the American Protestant churches in the early twentieth century were certainly opposed to evolutionism's implication that humans were derived from animals, which they interpreted as the cause of the contemporary decline in moral standards. Taking the Bible seriously on moral issues was also an integral part of their campaign. But—contrary the mythology that has built up around the Monkey Trial—there is no evidence that the early fundamentalists were united in taking up a literal interpretation of the Bible in general and of Genesis in particular. They were deeply concerned

by higher criticism, a branch of literary analysis that questions traditional claims about the composition and content of the Bible and
treats it as just another ancient text. But to hold that it was divinely
inspired did not entail taking every word literally, given that there
are many areas where it is clearly speaking in metaphorical terms.
Initially, only groups such as the Seventh Day Adventists insisted on
a literal reading of the text, and made this a central plank of their
opposition to evolutionism. The first phase of the fundamentalist
campaign was thus based primarily on concerns about the moral
implications of evolutionism. Only in the 1950s did biblical literalism become the foundation for mainstream creationism. Indeed,
widespread use of the term "creationism" begins with this transition (Numbers, 1992, 1998; for collections of primary sources see
Carpenter, ed., 1988 and Numbers, ed., 1994–95).

Some American evangelicals had remained suspicious of the idea
of evolution. But their opposition was muted, partly because they
had other concerns, and partly because at first there was little effort
to teach the subject in the schools. At the college level, the topic was
raised, and there is little evidence that professors were prevented
from teaching about evolution even in the southern states. The situation began to change in the early decades of the new century as
evolution theory began to filter into the school curriculum (on
changing images of science see Gilbert, 1997). At the same time
evangelicals became more concerned about the relaxation of moral
values, and hence more concerned to defend traditional beliefs. The
series of pamphlets known as *The Fundamentals,* issued between
1910 and 1915, were a rallying cry for the movement that became
known, appropriately, as fundamentalism.

Even now, though, opposition to evolutionism was not a central
plank in the campaign. There were two explicitly anti-evolutionist
pamphlets in *The Fundamentals,* but there were also several that
took a more relaxed line on the topic (Livingstone, 1987; Numbers,
1998). James Orr raised the issue in several contributions, stressing

that evolutionism did not require support for Darwin's theory of natural selection. As long as the process was supposed to be under divine guidance, there would be little to threaten the Christian's faith. There was certainly no need to take the text of Genesis literally on the details of how creation had unfolded. Other fundamentalists were less willing to compromise. One eminent defender of orthodox Christian doctrine, J. Graham Machen, compared the creation of the first human, Adam, with the immaculate conception of Christ—both required divine intervention to introduce something entirely new into the world (Machen, 1937).

The campaign to prevent evolutionism being taught in the schools began in earnest in 1921, when the popular Democratic politician and thrice-unsuccessful presidential candidate William Jennings Bryan gave a lecture, "The Menace of Evolution." Bryan was known as "the great commoner," and he articulated a growing distrust of the experts who were trying to direct society away from the cherished values of ordinary people. Where was the authority of the scientists who claimed to tell people that they were not competent to judge issues of vital concern to their daily lives? Bryan was deeply worried by the decline in contemporary moral values and blamed evolutionism in general, and Darwinism in particular, for teaching people to behave like animals. There was a widespread feeling that the Great War was a product of the Germans' adherence to a nationalist version of social Darwinism in which the strongest nation had the right to dominate those around it. The scientists themselves were aware of this point—it was articulated by the biologist Vernon Kellogg, who had talked with German officers during the war. Biologists were promoting non-Darwinian theories of evolution in part to block the claim that the theory necessarily endorsed the politics of "might is right." But the debate over the mechanism of evolution only fueled the suspicions of critics such as Bryan, since it gave the impression that evolution was anything but a clearly established scientific principle.

In 1922 Bryan threw his weight behind a campaign to ban the teaching of evolution in the public schools of Kentucky. The campaign flourished and was soon extended all over the country, although it was most active in the southern states. As Michael Ruse (2005: 153) points out, the South remained economically backward long after the Civil War, and people still looked to the Bible for reassurance that their sufferings were part of a divine plan. Neither they nor the impoverished factory workers in the North had benefited from the industrial progress that fueled the ideology of the Modernists, and both retained a more traditional faith. Presbyterians, Baptists, and Lutherans were most active in the anti-evolution crusade (Bryan himself was a Presbyterian), but many churches were divided on the issue. Campaigning was led by inter-denominational organizations such as the World's Christian Fundamentals Association. Paradoxically, many black religious leaders backed the fundamentalism, despite the link between figures such as Bryan and racism—they saw their own people as chosen by God (Moran, 2003, 2004). Over the next decade, twenty-three states would consider anti-evolution legislation. In the end only three actually enacted it—Arkansas, Mississippi, and Tennessee (although Oklahoma forbade the use of textbooks containing evolution and Florida formally condemned the theory).

Tennessee's Butler Act forbidding the teaching of evolution in the state's schools was introduced in January 1925 and passed into law six weeks later. The American Civil Liberties Union, an organization that campaigns for the freedom of speech, made it known that it would be willing to fund the defense in a test case. The leading citizens of the small town of Dayton realized that here might be a chance to get their community some publicity. They looked for a local teacher who would volunteer to challenge the law. John Thomas Scopes, a young science teacher, offered himself as a candidate and was duly charged. He taught physics and mathematics, not biology, and knew little about the details of evolution—but he accepted the theory in principle and was opposed to the legislation. The first steps toward one of the most highly publicized trials of the century had been taken. Scopes later wrote a personal account of the event (Scopes, 1967). Classic books on the trial are De Camp (1968), Ginger (1958), and Settle (1972); the best modern survey is Larson (1998). Larson's book exposes many of the myths that have grown up around the event, graphically portrayed in the movie *Inherit the Wind* of 1960, based on a 1955 play by Jerome Lawrence and Robert E. Lee (the title is from Proverbs, 11, 29: "He that troubleth his own house shall inherit the wind; and the fool shall be servant to the wise of heart.")

The leaders of the Dayton community wanted to turn the trial into a publicity circus, and this became a certainty when Bryan offered his services to the prosecution. The ACLU did not want a public debate and would have preferred to keep the trial focused on the constitutional status of the law. But their case also hit the headlines when the noted lawyer Clarence Darrow—a self-confessed agnostic and opponent of organized religion—jumped in with an offer to lead the defense team. The world's media descended on Dayton, led by H. L. Mencken of the *Baltimore Evening Sun,* who would orchestrate the efforts of the northern big-city papers to portray the South as a backward-looking relic of premodern society (on the publicity value of the images used in the trial see Clark, 2001).

The ACLU held public meetings in New York City ahead of the trial in which Scopes was introduced to Henry Fairfield Osborn and other notable scientists. Osborn was already campaigning on the evolution issue, promoting his theory that humans did not evolve directly from apes to check one of Bryan's most emotional arguments. He was the most prominent of the expert scientists that Darrow was hoping to call as a witness at the trial. In the end he did not attend, partly because he distrusted Darrow's hard-line atheism, but also because he had burnt his fingers over a very public misidentification of a fossil. Lesser scientific figures were called instead. Shailer Mathews, the leading Modernist theologian, would head a team of clergymen prepared to argue that evolution was acceptable to the Christian. Bryan too was attempting to line up expert witnesses, but the only man with scientific credentials he could find was out of the country. This was George McCready Price, one of the founders of young-earth creationism, who will figure prominently in our story. So Bryan and the prosecution team changed their tactics and decided to argue the case on strictly legal grounds: the people of Tennessee paid the teachers' salaries and had the right to monitor what they taught. In the end, the court ruled the testimony of all Darrow's experts to be inadmissible, although their statements were widely circulated in the press.

The trial was a scene of high drama, public excitement, and a good deal of acrimonious debate over side issues, such as whether or not it was appropriate to begin proceedings with a prayer. The technicalities were soon decided, making it clear that Scopes had indeed broken the law. Darrow then gave an impassioned speech outlining the defense's case that the law was unconstitutional because it violated guarantees on the freedom of speech. At the same time, he made clear his contempt for those who would impose outdated ideas on the next generation in the name of religion. Bryan too gave a speech in which he defended the right of ordinary people to make decisions on issues with moral and religious implications, including the creation of humankind. They also had the right to ensure that their deepest feelings were not violated by what was being taught to their children in publicly funded schools. The trial came to a head when Darrow, denied his expert witnesses, called Bryan himself to the stand as an expert on the Bible and its teachings. The resulting interchange is as much a part of the mythology of science as that between Thomas Henry Huxley and Bishop Wilberforce in 1860. Darrow effectively exposed the weakness of the literalist position on well-known stories such as Jonah and the whale. But—contrary to the popular image presented in many later accounts—Bryan did not defend the young-earth interpretation of Genesis. He had long been a supporter of the day-age theory in which each of the days of creation might be understood as a long period of time. Had he been able to call Price, the two would have disagreed on this point. At this moment in time only a very few creationists were prepared to see all the fossil-bearing rocks as relics of Noah's flood. The position that Price and his followers would eventually establish as mainstream creationism was as yet largely unheard.

There is little doubt that Bryan did his case little good and bitterly disappointed those in the courtroom, who expected a more vigorous defense of the scriptural position. The following exchange, toward the end of the cross-examination, gives a flavor of the tensions aroused. Asked what the purpose of his questions was, Darrow said, "We have the purpose of preventing bigots and ignoramuses from controlling the education of the United States," to which Bryan replied, "I am simply trying to protect the word of God against the greatest atheist or agnostic in the United States (prolonged applause). I want the papers to know I am not afraid to get on the stand in front of him and let him do his worst. I want the world to know" (prolonged applause) (from the court transcript, Anon, 1925: 299).

The classic image portrays Bryan—like Wilberforce—as a beaten man at the end of the debate. He died a few days later. But not everyone saw it that way, and even Mencken wrote that the fundamentalists had won the day. Scopes was indeed convicted and fined a hundred dollars (later rescinded on a technicality). The myth that the Monkey Trial was a defeat for the traditionalist forces only began to take shape in the following decade, and was eventually enshrined in the popular imagination in *Inherit the Wind*. According to the myth, the fundamentalists were exposed as country hicks out of touch with the modern world, and retreated into the hills. In fact, the anti-evolution campaign continued to be active for the rest of the decade and faded away only because it became clear that at the practical level it had been successful. If only three states passed anti-evolution legislation, the schools nevertheless stopped teaching the subject. To avoid confrontation, publishers ensured that the textbooks used in the nation's schools no longer contained references to evolution (Nelkin, 1977, 1983). For all intents and purposes, Bryan's campaign had succeeded. People would continue to read about evolution and debate its implications, but children would not be exposed to it in the schools.

The Monkey Trial was in many respects a transitional event. In some respects it symbolizes the start of what became known as the creationist movement, yet in others it marked the end of an era. Bryan and the opponents of Darwinism focused on the moral ambiguity of the theory of natural selection—yet most of the scientists involved were still promoting alternatives designed to evade the stigma of social Darwinism. Fundamentalism was a new and potent social force in America, but it was at loggerheads with a still-powerful liberal theology in the churches that made common cause with the anti-Darwinian evolutionists, who saw the ascent of life as necessarily progressive. As yet, the young-earth model of creationism was not the main alternative to evolutionism.

All this would change over the next few decades, and the creationist debates of the post–World War II decades would take place in a very different atmosphere. Darwinism was emerging from its eclipse in science, and would demand the right to be heard—and taught. But equally significantly, the Modernists found their influence in the churches undermined, even in Europe. World War I had struck a great blow against the idea of progress—how could Western culture be presented as the high-point of civilized development if it permitted this level of carnage? The Modernists were able to shrug this off in the 1920s, but the hectic culture of the jazz age in the big cities did not seem like progress to those outside the social elite. The economic depression of the 1930s and the rise of Fascism and Nazism in Europe drove home the message that there was something deeply flawed in the moral state of the West. Secularists turned to Marxism as a way of saving the idea of progress, but to many religious people it seemed that the liberals' optimistic hopes of perfecting humanity were misguided. Perhaps their efforts to free Christianity of its "outdated" traditions had missed the point. In their rush to portray humanity as the cutting edge of a divinely planned progressive trend, they had lost sight of the possibility that the whole scheme had become derailed.

This more traditional vision lay at the head of the movement known as neo-orthodoxy, which transformed the churches in the late 1930s and 1940s. Although superficially similar to fundamentalism, this was the product not of simple evangelical fervor, but of deep reflection by intellectuals. In Europe, the Swiss theologian Karl Barth called for a return to the traditional vision of humanity proclaimed in the Gospels: Human nature is deeply troubled because we have become alienated from God, and only His grace can save us. As Barth's influence gained ground in the Anglican church, Modernists such as Barnes and Raven found themselves increasingly marginalized (Bowler, 2001). Raven's vision of an immanent God as the source of nature's creativity meant nothing to theologians who took it for granted that the material world was merely a backdrop to the spiritual drama of humanity's fall. Neo-orthodoxy didn't want an alternative view of creation, or a return to the argument from design—it just wasn't interested in science. C. S. Lewis, who wrote popular religious works as well as his better-known childrens' stories, seems to have dismissed evolutionism as so obviously misguided that it was hardly worth arguing against.

In America the same points were being raised by theologians such as Reinhold Niebuhr, a former pastor who taught at the Union Theologial Seminary in New York. His philosophy of Christian realism was an open challenge to the optimism of Protestant liberalism and to the loosened social values of the time. Although not endorsing the complete pessimism of traditional orthodoxy, Niebuhr saw evil as a product of humanity's refusal to acknowledge God. He held out no hope of social justice in this world and insisted on the tragic nature of the human predicament. Like Barth and his followers, he simply had no interest in science. Where the fundamentalists had seen the struggle against evolution as a major part of their campaign to preserve traditional Christian values, Christian realism subverted Modernism from within by undermining the intellectual foundations of belief in progress.

In the postwar years, the liberal approach to Christianity would have to struggle hard to retain its credibility among the more theologically sophisticated members of the churches. But if it tried to re-establish a link with science and with the idea of evolution, it would confront a resurgent Darwinism that had little room for mysterious progressive trends or built-in goals. Meanwhile, in America at least, fundamentalism was gaining influence decade by decade, and it was the more strident anti-evolutionists who were increasingly making the running. The next phase of the debate would be fought over very different ground to that disputed during the Monkey Trial.

CHAPTER FIVE

MODERN DEBATES

The 1940s saw a consolidation of the synthesis between genetics and the Darwinian selection theory. Julian Huxley's classic *Evolution: The Modern Synthesis* was published in 1942, when the Second World War still hung in the balance. In the postwar decades it became clear that science would no longer tolerate the vaguely teleological theories of evolutionary progress which had flourished during the eclipse of Darwinism. The world would have to deal with the consequences of a fully materialistic theory of evolution, and increasingly there were radicals such as Richard Dawkins who would insist that such a theory made nonsense out of any form of religious belief. Humans are not the intended products of a divinely inspired progressive trend. We are just lumbering robots programmed by selfish genes. In Europe, where the trend of secularization continued apace, such views were resisted more for their ideological consequences than because they disturbed religious belief. But religion was still a powerful force in America, the ideal bulwark against the godless Marxism confronted during the Cold War. And what really disturbed religious traditionalists in America was that the implications of Darwinism were no longer confined to the intellectual world. Inspired by a new level of confidence, the Darwinists demanded the kind of access to the schools that they had more or less voluntarily abandoned in the aftermath of the Monkey Trial.

The compromise that had deflected the original fundamentalist attack had broken down. The backlash was not slow to emerge. Religious fundamentalists renewed their campaign against evolutionism and focused their attention on resisting its influence in the schools. But there was something new about this postwar campaign. Increasingly, the impetus behind the anti-Darwinian movement was provided by an extreme form of biblical literalism, the movement we now call young-earth creationism. In the era of the Monkey Trial, only a few highly conservative sects had tried to insist that the whole edifice of the scientific view of earth history was flawed. Now the claim that the earth is only a few thousand years old emerged as a powerful alternative among those whose real concern was defending the truth of Genesis. For these religious conservatives, the best way of sustaining the view that salvation can only come through the second coming of Christ was to deny not only evolutionism, but also the geological and paleontological foundations upon which the theory rested.

The campaign for equal time to be given to what was called "creation science" in the schools eventually foundered on the ACLU's ability to show that its tenets were inspired by Genesis and not derived from any reasonable interpretation of the evidence. If creation science was fundamentalist Christianity in disguise, then the first amendment to the Constitution—designed to ensure the separation of church and state—forbids its teaching in the schools. The later movement known as Intelligent Design (ID), which focuses instead on a modernized version of Paley's argument from design, was introduced to bypass this problem, and its campaign continues unabated today. Much of the public support for ID still comes, however, from conservatives whose real position is based on the young-earth interpretation of Genesis. Meanwhile, creationism spreads around the world. Evangelical sects are active in Africa,

South America, and even in Europe (which they see as being as much in need of missionary activity as the rest of the world was in former centuries). Islam too has its fundamentalists, as we are all now aware, and these have their own reasons for opposing an evolution theory that denies the creation stories embedded in a very different sacred text.

Here is the basis for the widespread opinion that evolutionism and religion must by their very natures be in a state of conflict. Yet if there is one message to be derived from the previous chapters of this book, it is that such a polarized debate has few historical antecedents. In the hundred and fifty years between 1800 and 1950, hardly any educated person would have endorsed the position we now call young-earth creationism. The fact that the earth had changed over a long period of time was accepted even by those who found the theory of evolution disturbing. And within the evolutionists' camp there were many whose liberal views on religion allowed them to search for a way of regarding evolution as the unfolding of a divine plan established by a God whose activity is immanent within the universe. This middle ground has not disappeared in the modern world, and although it may not get the same degree of publicity as the big confrontations, it is still the scene of considerable activity.

In the immediate postwar years, there were some religious thinkers who still seemed to think it might be possible to retain the kind of teleological evolutionism permitted by the old non-Darwinian theories. They were soon disabused of this hope by the critical attitude of the scientific community. At a more sophisticated level, liberal religious thinkers talk of a "process theology," which seeks an alternative to materialism through the philosophies of figures such as Alfred North Whitehead. This approach may have its academic supporters, but it says little about the debates in biology, which center on the selection theory and the notion of the selfish gene. Those who seek to resist the materialism of Darwinists such as Dawkins without slipping into outright appeals to the supernatural are active at a number of levels. Perhaps an "open-ended" form of evolution is the only way that God could create beings with free will. Curiously, it has begun to emerge that it may be precisely in the pessimism of the Christian message, its emphasis on suffering as a key ingredient in both the human and divine situations, which might allow a reconciliation with a scientific theory that is built on the model of the "struggle for existence." The liberal Christian tradition remains an active participant in the debate concerning human origins.

THE NEW DARWINISM

The modern Darwinian synthesis, which has dominated biological thought since the 1950s, had its origins in the genetical theory of natural selection put together by R. A. Fisher, J. B. S. Haldane, and Sewell Wright during the interwar years. When brought together with the insights of the field naturalists and paleontologists, the new Darwinism provided a powerful model that was increasingly seen as providing a complete explanation of the evolutionary process. Since it was based on an updated form of Darwin's original theory, the new synthesis had the potential to undermine most aspects of traditional natural theology. There was nothing capable of providing a trend toward a predictable final goal, and no guarantee that all or even most evolution would be progressive in any meaningful sense. Evolution was a process that adapted populations to their local environment, and this process worked by something very like Darwin's original model based on random variation (now identified as genetic mutations) and survival in the struggle for existence.

All the characters of the human mind had to be a product of this process applied to the specific circumstances in which our ancestors had diverged from those of the great apes after they had moved out onto the African savannah. By the 1950s it was clear that the Australopithecines, first discovered in South Africa in the 1920s, were the earliest members of the human family. They had stood upright, but had brains scarcely larger than an ape's. The great expansion of the human brain came later, and was interrupted by long periods of stagnation. Eventually a species known as *Homo erectus* (Dubois's old "Java man"), armed with fire and stone tools, spread out of Africa over the whole of the Old World. Modern humans are a very late offshoot of the same stem, also arising first in Africa (Reader, 1981; Walker and Shipman, 1996). Scientists still argue over the factors that triggered the key steps, but the steps themselves are clear enough. There is little sign of a goal-directed trend forcing us to progress toward higher levels of mental or moral awareness.

In the hands of a later generation of atheistic Darwinists such as Richard Dawkins and Daniel Dennett, all these implications would be unpacked with a relish that only confirmed the instinctive fears of many religious believers. But as we have already noted in the previous chapter, there was at first no indication that the theory would have to be linked to a total rejection of purpose in nature. A surprising number of the founders of modern Darwinism were influenced in their youth by progressionist visions of evolution such as Bergson's theory of the creative *élan vital*. Lloyd Morgan's theory of emergent evolution also allowed them to believe that new levels of activity such as the human mind could appear at key points in the advance of life. Some, including R. A. Fisher and Theodosius Dobzhansky, retained a liberal form of Christian faith. Others retained their commitment to the ideology of progress even when they abandoned any belief in formal religion (Ruse, 1996). Although reluctant to concede that there was a personal God who intended the process to generate human beings, they continued to hope that Darwinism could be linked to some more general idea that human values are not mere accidents, but are products of the laws which govern the cosmos. The new Darwinism may have undermined the

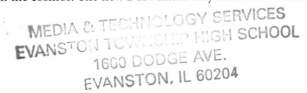

more explicitly teleological aspects of the old non-Darwinian theories, but it was not at first presented as something that would necessitate rejection of all forms of religious faith.

Fisher remained a practicing Anglican throughout his career, occasionally preaching in his college chapel at Cambridge. He continued to see natural selection as a form of "creative evolution." In a lecture delivered in 1950 he argued that, contrary to popular belief, it was Darwinism rather than Lamarckism which could more easily be reconciled with Christianity. Lamarckism suggested that organisms could influence the future by striving to achieve a goal, but for the Darwinist it was the actual outcome of the activity that mattered—it was a case of doing or dying. This, he claimed, offered a close parallel to the Christian belief that God requires us to perform good works, rather than expecting "good intentions and pious observances" (Fisher, 1950: 19–20).

Julian Huxley, whose 1942 book provided the new Darwinism with its name, "the modern synthesis," was also influenced by Bergson. He remained committed to the view that evolution is progressive, and that humans are the inevitable goal toward which it has been advancing (Greene, 1990; Huxley, 1970; Waters and van Helden, 1992; Swetlitz, 1995). He had abandoned belief in a personal God before the war, but still argued that the emotions that lay behind religious belief were an important aspect of what makes us human. He sought a "religion without revelation" in which people could still see themselves as playing a purposeful role in the cosmic drama of progress. He now emerged as a leading spokesman for the philosophy of humanism, which sought to redefine our duties and aspirations in purely human terms, without any appeal to the supernatural (Huxley, 1961, 1964). But in many ways his thinking still paralleled that of the liberal religious believers with whom he remained on good terms. His support for Darwinism certainly did not force him to accept that the universe is without purpose, or that human life is essentially meaningless. In 1959 he provided a very positive introduction to the translation of Teilhard de Chardin's *The Phenomenon of Man*, a belated expression of the kind of teleological, non-Darwinian evolutionism that most biologists now thought their science had ruled out. We shall return to Teilhard later in this chapter, but at this point it is important to note that Huxley's support indicates a degree of ambivalence about the extent to which the new selection theory had really destroyed the liberal Christian faith in progress.

Some of the American founders of the modern Darwinian synthesis displayed a similar lack of enthusiasm for materialism. Most notably, the Russian emigré Theodosius Dobzhansky remained a lifelong member of the Eastern Orthodox Church. Dobzhansky's *Genetics and the Origin of Species* of 1937 was one of the key texts helping to link the genetics of natural selection with field studies (see Adams, 1994). He recognized the danger that some church leaders might encourage a rejection of scientific advances, but saw no barrier himself between his work as an evolutionist and a liberal Christian faith. In fact, for Dobzhansky, Christianity was an evolutionary religion, since it called for everyone to participate in future human progress. As a refugee from the Soviets, he was particularly concerned to stress the role of human freedom as the essential basis for such participation. His 1967 book *The Biology of Ultimate Concern* made these commitments in a very explicit way. Like Huxley, he was also attracted to Teilhard de Chardin's evolutionary mysticism. He accepted the need to see evolution rising steadily to ever-higher levels of mental and spiritual development, although as a good Darwinian he found it difficult to accept that the specific character of the human species was predetermined. Evolution groped its way forward in an uncertain and unpredictable manner—it did not run straight toward a fixed goal.

This last point was also central to the thinking of George Gaylord Simpson, who did most to bring paleontology within the sphere of the new Darwinism. Simpson was raised by a fundamentalist Presbyterian family and lost his faith while at college. But he always felt the need for an emotionally satisfying replacement, and often worshipped with the Unitarians. In a series of essays (collected in Simpson, 1963) he expressed his belief in the progressive nature of evolution and in the significance of human individuality as the high point reached by the process here on earth. But like Dobzhansky, he recognized that the unpredictable nature of the Darwinian process meant that even if life existed on other worlds, the chances of it evolving into something exactly resembling humanity were remote. He knew Teilhard and appreciated the emotional power of his vision, but he could not accept that humans were somehow the intended products of progress. Darwinism simply didn't allow for this kind of direction to be imposed on the process. It was hard for many religious people to accept such a model of evolution—it implied that even a slight change in the circumstances affecting our ancestry at any point in the past might have led to a world in which there was nothing resembling human beings. This was hard to reconcile with the belief that we are central to the Creator's purpose.

Even in later decades there are biologists who have made major contributions to the development of Darwinism, but have retained a respect for religious belief. Edward O. Wilson is the pioneer of sociobiology, the application of a very individualistic model of natural selection to explain the instincts that govern social behavior. Using the idea of what Richard Dawkins would call the "selfish gene," sociobiology has been immensely successful in explaining the behavior of many animal species, including Wilson's favorites, the ants. But when Wilson applied the model to human behavior in the final chapter of his 1975 text *Sociobiology,* the result was uproar (Caplan, 1978). His claim that at least some aspects of our behavior are controlled by instincts programmed into us during the last phases of our evolution was greeted with outrage by the social sciences and by those on the political left. It also undermined exactly the point made by Dobzhansky, that evolution had led to an increase in human freedom. Wilson subsequently refined his model (1978) but without backing away from the basic idea.

Like Simpson, Wilson was raised a fundamentalist—he was "born again" at the age of fifteen and lost his faith at college. But he too has found it hard to shake off the view that religion plays an important role in our lives. Wilson has aroused the ire of churchgoers by suggesting that we are programmed to accept beliefs which help to cement our social group, and that religion flourishes because of this instinct. But he shares the vision of progress accepted by many liberal Christians, and also promotes the view that our sense of duty—the foundation of morality—is built into us by our biological nature. In another direction entirely, Wilson has emerged as the proponent of what could be counted as a new natural theology. He has celebrated the diversity of life produced by evolution, and has endorsed environmentalists' concerns about the extinctions produced as a consequence of human activity. He writes: "Those committed by religion to believe that life was put on earth in one divine stroke will recognize that we are destroying the Creation, and those who perceive biodiversity to be the product of blind evolution will agree" (1992: 335). Given that the fundamentalists actually have a poor record on environmental issues, Wilson's willingness to see all living things as a precious part of our heritage offers a worldview which for some may be as important as conventional religion.

Wilson's Darwinism may have been incorporated into a moral vision that in some ways parallels the functions of religion. But to many Christians his views revive all the dangers that have long been identified as consequences of the selection theory. There can be little doubt that as modern Darwinism has triumphed within science, more of its exponents have been encouraged to present it as the foundation for an outright attack on religion. The dangers were recognized at a very early stage by David Lack, a student of Julian Huxley. Lack's 1947 book on the Galapagos islands, *Darwin's*

Finches, coined the popular name for the birds whose geographical diversity is seen as a key piece of evidence for evolution. In the same year, Lack joined the Anglican church, and ten years later he published a book identifying the tensions he now saw between his Darwinism and his Christian faith. Entitled *Evolutionary Theory and Christian Belief: The Unresolved Conflict,* the book pinpointed several key areas of concern. Lack knew that no modern biologist could accept the old idea of a creative life force, and although he acknowledged that Fisher and others saw natural selection as a creative process, he thought there were still major difficulties in the way of reconciling a process based on undirected mutations with the belief that nature reflects a divine purpose. But it was human nature, not design, that most bothered Lack. He thought it impossible for natural selection to promote moral instincts, and he believed that there was something unique about the human soul. No evolutionist could accept a literal reading of the Genesis story of the Fall, but our sense of responsibility was a new force in the world and the Christian needed to believe that this force had a divine origin.

Lack's fears have been fully justified by later events, as an increasing number of Darwinists have sought to exploit the theory as ammunition in a war against religion. Of the founders of the modern synthesis, J. B. S. Haldane had always been a critic of Christian belief, and in the 1930s he had become a Marxist (he died in India, where he had moved as a protest against British imperialism). In 1971 the French biologist Jacques Monod created a stir with his book, *Chance and Necessity,* in which the whole idea of purpose in the universe was dismissed. Given "chance" (i.e., undirected) variation within populations, then the necessity of survival ensures that the best-adapted characters will take over and shape the production of new characters. This materialist strand of thought since developed apace, and two figures have emerged as the champions of atheistic Darwinism. The British biologist Richard Dawkins is a brilliant popular science writer who has put his skills at the service of the campaign to replace traditional views of nature and human nature with the Darwinian perspective. And the American philosopher Daniel Dennett has attracted the same level of praise and criticism with his efforts to apply the Darwinian approach, especially to the origins of the human mind.

Dawkins made his name with an important exposition of the new Darwinism, *The Selfish Gene* (1976). The term "selfish gene" has come to encapsulate the logic of natural selection when visualized from a genetic perspective. When applied to behavioral instincts, it underpins the arguments of sociobiology. It shows us that we need to move beyond Darwin's emphasis on the success or failure of individual organisms to think in terms of the success or failure of genes in the process that determines how they replicate. Of course genes aren't selfish and they don't have a point of view— they are just bits of DNA that can copy themselves—but Dawkins realized how powerful the metaphor of purpose is for helping people to visualize what is going on. From the gene's point of view, it may be better for you to sacrifice yourself to save the lives of your relatives, because they carry some of the same genes, and if they survive, more copies might make it to the next generation than if you had reproduced yourself. The genes could thus program you with a potentially fatal altruistic instinct. They would certainly program you to be blindly indifferent to the suffering of other organisms you depend on for food.

Dawkins thus portrays animals as lumbering robots programmed to behave in certain ways by their genes, the programs themselves being shaped solely by the pitiless logic of success in the race to duplicate. Unlike Wilson, though, he was more willing to see humans as the one species that has developed the capacity to escape the tyranny of the genes. Because we can innovate and learn, we have been able to develop cultures that seem to violate some of the most basic constraints imposed by natural selection. However, Dawkins has extended the Darwinian model into the world of culture by postulating that ideas themselves can be treated as replicating entities which seek to spread into and dominate our culture. These are "memes," and their competition defines much of our social and cultural life. We are bombarded with new ideas seeking to influence how we behave, of which only a few are successful. But those that do "catch on" come to affect the habits and attitudes of almost everyone—and may then mutate to produce variants which compete to affect the lives of future generations. Significantly, religious beliefs are memes, and their success can thus be explained by a Darwinian process operating at the cultural rather than the biological level.

The Selfish Gene was not an overtly anti-religious book, but the logic of its vision of a totally Darwinian universe was obvious enough. Why does the world contain parasites that cause immense suffering in their prey? This is a puzzle that has baffled natural theologians and played an important role in Darwin's own loss of faith. But for Dawkins it is inevitable that an evolutionary process driven solely by selfish genes will contain such organisms. The concept of a benevolent God isn't just superfluous—it is patently incapable of explaining the kind of universe we live in. Crucially, for Dawkins, natural selection can explain the one thing that the natural theologians had always insisted was of supernatural origin: complexity. His books The Blind Watchmaker (1986) and Climbing Mount Improbable (1996) tackle the argument from design head on. The Blind Watchmaker is, of course, a reference to Paley's argument of the watch and the watchmaker: when we see the complexity of the watch, every piece adapted for the purpose of constructing a machine to tell the time, we know that the only way it could have been designed and built was with the intelligence of the watchmaker.

Paley thought animals were in the same position as watches, but Darwin introduced the idea of natural selection precisely to show that the everyday laws of variation and survival could produce the same results, provided they could be seen to operate over vast periods of time. Dawkins's accounts of the modern version of the selection theory effectively demolished many of the oversimplified objections that have become the creationists' stock-in-trade, and did so in a language which everyone could understand. The old "monkeys at the typewriter" argument crumbles before his demonstration that if you reiterate a random process over many generations, saving the best results each time, eventually you will get a line from Shakespeare out of the initial gibberish—even though the chance of success in any one attempt is vanishingly small.

Put this demolition of design together with the demonstration of the harshness of the world we live in, and you have a message that—to Dawkins—tolls the death knell of religion. The idea of a designing God isn't just superfluous, it's positively wrong: "In a universe of blind physical forces and genetic replication, some people are going to get hurt, other people are going to get lucky, and you won't find any rhyme or reason in it, nor any justice. The universe we observe has precisely the properties we should expect if there is, at bottom, no design, no purpose, no evil and no good, nothing but blind, pitiless indifference" (Dawkins, 1995:153). Dawkins believes that Darwinism makes it possible for someone to be "an intellectually fulfilled atheist." In later years he has become not only a vocal proponent of atheism, but an outright opponent of organized religion. He now believes that (because they are memes in conflict with one another) the world's religions are a positive danger to humanity (Dawkins, 2006). They automatically breed dogmatism and intolerance, and have once again become the most active driving force of conflict, war, and terrorism. Critics from a liberal theological perspective decry Dawkins's oversimplified vision of religious belief (e.g., McGrath, 2005). But I suspect that he would respond by asking how many of the world's religious believers belong to the intellectual elite that preaches tolerance—and how many to the extremist sects convinced that their beliefs alone offer the sole route to God?

In America the philosopher Daniel Dennett has taken a role parallel to that of Dawkins, hailing Darwin's theory of natural selection as perhaps the most important idea ever. It will allow us to dispense once and for all with the notion of a God who designed the world. Dennett focuses not on the biology of Darwinism but on the logical structure of the theory of natural selection. His book *Darwin's Dangerous Idea* (1995) argued that the combination of random variation with selection could be applied across a wide spectrum of natural processes, not just in biology, to explain the emergence of complexity. Natural selection is an algorithmic process, an endlessly repeated sequence of individually mindless steps that is guaranteed to generate improvements over time. Because it can be seen operating in virtually every natural process, it becomes a "universal acid" that eats away the foundations of traditional beliefs. Wherever we once thought we saw evidence of a designing Mind at work, we can now understand how the same results have been produced by the mindless operations of natural selection.

In particular, Dennett has applied Darwinism to explain the development of the human mind and of culture. He uses new approaches in evolutionary psychology, some of which see the brain's learning capacity as a process of trial and error, to account for human consciousness. In essence, the human mind is no different from the artificial intelligences that are produced by computer programmers—we just find it difficult to appreciate that the ideas and sensations which appear in our consciousness have actually been generated by mechanistic processes in the brain. Like Dawkins, Dennett sees sociobiology as the best way of understanding how our minds have been programmed to accept certain modes of behavior as natural. He also sees culture as an arena of competing memes, each seeking to shape the behavior of as many people as possible. Significantly, in recent years Dennett too has begun to stress the harmful effects of religion and to suggest that the corresponding memes should be rooted out from the foundations of Western culture (Dennett, 2006). We must create our own values, not rely on rigid prescriptions passed down by an imaginary God.

There have been some eminent evolutionists who have disagreed with what they see as the Darwinian fundamentalism promoted by Dawkins and Dennett. The late Stephen Jay Gould argued long and hard against the claim that the natural selection of small genetic variations could explain the whole development of life on earth. According to Gould, evolution is constrained by the processes of embryological development which unpack the genetic information, so that variation is often far from random. The modern science of evolutionary developmental biology, popularly known as "evo-devo," promotes these ideas and is making significant changes to our vision of how evolution has unfolded (Carroll, 2005). Gould was also hostile to the atheism promoted by the hard-line Darwinists, arguing that science and religion exist in different worlds and cannot really affect each other (Gould, 1999). Even so, in some respects Gould was a fully fledged Darwinian in a way that would upset any Christian traditionalist. In his account of the famous Burgess shale fossils (1989) he argued that these bizarre relics of some of the earliest known animals suggest that evolution could easily have developed in entirely different directions to those it has actually taken. Because every transformation depends on the haphazardly changing local environment, if we could "replay the tape" of life, the outcome would be very different. George Gaylord Simpson's claim that humans are not the predictable outcome of evolution is thus vindicated.

THE RESURGENCE OF CREATIONISM

Following the Monkey Trial, fundamentalist opposition to evolutionism gradually died down, in part because the subject did indeed get dropped from the school curriculum in most states. But in the 1950s and 1960s, supporters of the new Darwinian synthesis became increasingly anxious to see their science take its place in education. As the geneticist Hermann Muller wrote on the centenary of the *Origin of Species,* "One hundred years without Darwin are enough" (Muller, 1959). The campaign was helped by the enormous pressure to improve the country's standing in science following the success of the Soviet Union in launching the first satellite, Sputnik, in 1957. The re-entry of evolutionism into the schools horrified the fundamentalists and prompted them to renew their opposition to it. From the 1960s onward, America has been the scene of a war between fundamentalist religion and the orthodox scientific community that has passed through a number of different phases and shows no signs of diminishing. In recent years evangelical Christian groups have also begun to have some success in promoting creationism in other countries. Meanwhile, as many Islamic nations struggle to gain access to modern science in order to exploit its technological spin-offs, Islamic fundamentalism has begun to resist the spread of what it too regards as an atheistic theory of evolution.

Although the anti-evolution campaign seems to present a united front against the hated theory, it is in fact a loose coalition of groups with significantly different positions. At the time of the Monkey Trial, few fundamentalists argued for what we now call the "young-earth" position, which postulates that the creation took place only a few thousand years ago and that all the fossil-bearing rocks were laid down in Noah's flood. Genesis was widely interpreted as allowing for a long period of time and multiple creations before the present one, either through the gap theory (in which there is a long period of time between the events in the first and the second verses of Genesis I) or the day-age theory (in which each day of creation is interpreted as a vast period of time, in effect as a geological age). These two positions are still accepted by a significant number of fundamentalists. But in the 1960s the young-earth position was revived and soon became the mainstay of creation science. Indeed the very word "creationism" came to be understood as referring to this particular alternative.

Supporters of the rival positions are quite capable of falling out with each other in a dramatic fashion, and these disagreements often reflect the very different theological positions adopted by the various Protestant denominations. The churches favoring a premillenarian vision prefer the young-earth approach because this vindicates a literal interpretation of the Bible and their expectation of the imminent end of the present world. Many Lutheran and Calvinist churches also support young-earth creationism. Baptists, however, are divided between fundamentalist and more liberal interpretations. But the debates between the rival Christian factions get little press attention because the main focus is always on the negative side of the campaign, the attack on evolutionism.

It is also worth noting that creationism has not been the only source of opposition to the theory of evolution. The counterculture of the 1960s spawned a number of movements opposed to orthodox science. Some of these took up positions whose negative aspects are identical to creationism, although the alternatives they offer to evolution are very different. Immanuel Velikovsky's book *Worlds in Collision* (1950) eventually gained cult status by convincing many that mainstream science was covering up evidence that would demolish the accepted story of the earth's history (De Grazia, 1966; Goldsmith, 1977). Velikovsky linked biblical stories of geological catastrophes with those derived from other cultures and reintroduced the catastrophist theory that had been popular in the early nineteenth century. But for all that he took the ancient texts seriously, Velilovsky's explanations were strictly naturalistic (if highly fanciful), depending on near collisions between planets and rapid evolution produced by massive bursts of mutation triggered by radiation from space. Erich von Däniken's *Chariots of the Gods?* (1970) also became a cult classic with its supposed "evidence" for early humans having been given technological help by extraterrestrials. In a later book (1977) von Däniken attacked the theory of evolution and claimed that the sudden appearance of new animal types at key points in the history of life was due to genetic engineering by the aliens.

At one level, the supporters of Velikovsky and von Däniken had a different agenda from that of the creationists—their suspicion of the scientific community was certainly not driven by a commitment to an alternative vision derived from revelation. But all of these movements shared a common hostility to the power of "experts," who were portrayed as imposing a rigid orthodoxy intolerant of dissent. It has been argued that creationism itself can be understood as part of this general wave of dissatisfaction with the elitism and alleged dogmatism of orthodox science (Nelkin, 1977). Furthermore, the arguments used against Darwinism by Velikovsky and von Däniken were often identical to those employed by creationists, although the supporters of the rival alternatives seldom bothered to debate among themselves. They were far more interested in trying to discredit the orthodox scientific position, and each movement simply assumed that if that could be done, its own position would be recognized as the most obvious replacement.

Whatever the initial parallels, support for Velikovsky and von Däniken has waned, while creationism has gone from strength to strength. Something more is at work here, and that something must be explained in terms of religious fundamentalism's offer of an alternative not just to science, but to the whole direction of modern life. Indeed, there is a sense in which the creationists are not hostile to science as such, as long as it stays in its place. They are happy to use the new technologies made possible by science, and—unlike the counterculture of previous decades—their ranks are increasingly filled by prosperous, urban Americans who are by no means hostile to the military-industrial complex. Science is only perceived as a threat when it goes beyond the search for means to control nature to investigate the origins of the structures that make up the universe as we know it. If it then seeks to undermine the creation stories upon which traditional Christian values are founded, it oversteps its bounds. Yet to many scientists, the freedom to erect theories that challenge traditional myths is integral to the whole project by which human reason seeks to understand the world. If that means reinterpreting the stories told in the Bible, so be it. Creationism works because so many people see their commitment to the Bible as both a source of salvation and a way of preserving traditional American values. This is why the biblical literalism of young-earth creationism has become a dominant force in American society without undermining support for science as a practical activity linked to technology and medicine.

In the previous chapter we noted that in the interwar years, there was one figure who stands out for his promotion of the young-earth position and flood geology, the Seventh-Day Adventist George McCready Price (on the history of creationism see Numbers, 1992, 1998, 2006, and for a collection of sources see Numbers, ed., 1994–95). In 1937 Price joined other Adventists to found a short-lived Deluge Geology Society. This was shunned by the main evangelical group addressing scientific issues, the American Scientific Affiliation, which endorsed an old-earth view of history with either multiple creations or even theistic (divinely guided) evolution. In 1954 Bernard Ramm's widely promoted *The Christian View of Science and Scripture* supported the gap theory to reconcile Genesis with old-earth geology. But the amount of attention paid to this book annoyed the small number of young-earth creationists, including John C. Whitcomb, Jr., and the engineer Henry M. Morris, who would take up a position at the Virginia Polytechnic Institute in 1957. Price himself was now getting old, although Whitcomb at first offered to collaborate with him and eventually produced a manuscript on flood geology that made numerous references to Price's work (and to Velikovsky). Eventually Whitcomb teamed up with Morris to write a revised book that deliberately sought to distance itself from Price and the Adventists. After the original publisher pulled out because the book was too literalist, this finally appeared as *The Genesis Flood* (Whitcomb and Morris, 1961).

The Genesis Flood would become the cornerstone of young-earth creationism, eventually selling over 200,000 copies. The book is not just an attack on evolutionism: because it endorses flood geology it presents an alternative to the whole orthodox scientific view of earth history. It thus rejects the established theoretical positions of geology, paleontology, and prehistoric archaeology. Because its dating of the creation gives a timescale shorter than some radiocarbon dates, even the physics used to uphold that dating technique is questioned. In a sense, there would be scarcely any need to argue against Darwinism from this background—any form of evolutionism just drops out as a byproduct of limiting the age of the earth and denying the extent of geological time. Whitcomb and Morris argued against the reality of the stratigraphical sequence used by geologists by pointing to a few areas where supposedly older rocks lie on top of younger ones. These are usually explained as "overthrusts"—blocks that have been pushed horizontally on top of neighboring rocks by the movements of the earth's crust which generate continental drift. But for the flood geologists they are an indication that the orthodox sequence is not universal. They explain the overall succession of strata by invoking differential sorting of the debris—including organic remains—settling out in the waters of the great flood. Currents within the flood waters explain why the normal sequence of deposition is sometimes interrupted. The first edition of the book referred to alleged cases in which giant human footprints had been found alongside those of dinosaurs. These were never substantiated and were dropped from later editions, although they are still often referred to in other creationist literature. Whitcomb and Morris appealed to Genesis to argue that the waters of the flood were derived from a canopy of water which once surrounded the earth. This protected the early surface from cosmic rays, accounting for the long lives of the patriarchs and the exaggerated ages given by radiocarbon dating. They took the Noah's Ark story literally, although they held out little hope of the Ark's remains being found. This has not prevented the circulation of numerous accounts of the Ark being located on Mount Ararat. Some creationists have gone to great lengths to work out how enough animals can have been carried in the Ark.

The Genesis Flood aroused controversy in evangelical circles, but it also inspired a new generation of young-earth creationists to begin spreading their message. Morris himself campaigned aggressively, to the increasing embarrassment of Virginia Polytechnic Institute. Whitcomb linked up with geneticist Walter E. Lammerts to establish an anti-evolution group within the American Scientific Affiliation. Along with Berkeley-trained biochemist Duane T. Gish and others they formed the Creation Research Society in 1963. This soon evolved into a dedicated pressure group for young-earth creationism. A textbook was eventually published to present the creationist approach to the life sciences (Moore and Slusher, eds., 1970). In 1970 Morris and others founded the Creation Science Research Center in San Diego, although this soon fragmented and Morris joined forces with Gish to found the Institute for Creation Research. Meanwhile the Adventists had also created an organization to promote flood geology, the Geosciences Research Institute at Loma Linda.

The various creationist institutions now became active on a number of fronts. To the dismay of the orthodox scientific community they began to challenge its authority both in public and in the schools. They issued a flood of pamphlets and organized lecture tours—often focused on college campuses—in which scientists were forced to defend their position against skilled debaters. Duane T. Gish emerged as perhaps the most challenging exponent of the creationist position. He forced the biologists onto the defensive by focusing on apparent problems that could be exposed in the public's oversimplified image of evolution (Gish, 1972). The discontinuity of the fossil record was a favorite area of debate, with the scientists forced to respond by offering what looked like complicated excuses to explain why the evidence for evolution was not as strong as people had thought. The same tactic also worked against natural selection: it was easy to ridicule the idea that complex structures could be built up by random mutations. It was as likely as a car emerging from a crash with an improved engine and transmission. To defend the Darwinian position, the scientist had to delve into technicalities and risked boring the audience or appearing aloof and condescending. By setting the agenda for each debate, the creationists were able to focus on those areas where they could put the evolutionists in a defensive position, leaving the audience with the impression that the scientists were covering up major deficiencies in their position. All the attention was focused on attacking evolution, and there was often little effort to explain the details of the young-earth alternative. The evolutionist was thus given no opportunity to challenge the credibility of flood geology.

Equally serious was the campaign to have creation science included in the school curriculum of American states. In the 1960s any remaining state laws forbidding the teaching of evolutionism were declared unconstitutional. The National Science Foundation began a campaign to modernize the science curriculum in schools across the nation, with evolution as a prominent feature. The new generation of creationists now focused on an attempt to require states to include "equal time" in school science lessons for the creationist alternative (Nelkin, 1977, 1983). There were extensive debates in California during the early 1970s. Creation science was eventually excluded, but only after the wording of textbooks was watered down to make evolutionism less visible. In 1972 both the National Academy of Sciences and the American Association for the Advancement of Science issued statements condemning creation science.

As in the public lectures, much of the so-called evidence that would have been included in the teaching of creation science consisted of attacks on an oversimplified model of evolution theory. But behind this lay the young-earth movement's efforts to replace the orthodox view of the earth's history with flood geology. Once this could be exposed, it was possible to argue that creation science was an attempt to uphold a literal interpretation of the Genesis story. The American Civil Liberties Union was then able to argue that creation science was religion, not real science, and could not be taught in the schools. The dispute came to a head in 1983, when an Arkansas law requiring equal time was struck down after a trial that hit the headlines all over the country. Many experts testified for the evolutionists, and the philosopher Michael Ruse was called in to undermine the scientific credentials of the creationists. He showed how the creationists used an oversimplified view of the scientific method to dismiss evolution as "only a theory" while concealing their unwillingness to expose their own alternative to rigorous testing.

It now became clear to the creationist movement that, whatever the level of support for creation science in their own ranks, the young-earth position was an obstacle to their hopes of getting an alternative to evolutionism into the schools. Into the breach now stepped a new form of creationism, still intent on exposing the weakness of evolution theory, but now based on a revival of the old argument from design. This is the theory of Intelligent Design (ID), which has become the focus for a new wave of anti-evolution legislation. To the dismay of young-earth creationists, the supporters of ID make no effort to construct a detailed history of life based on supernatural events. Some are even willing to admit large amounts of evolution in the later development of life. They limit themselves to demonstrating the inability of orthodox Darwinism to explain the complexity of living things. If Dawkins wants to replace Paley's watchmaker God with natural selection, ID wants to shift the focus of debate into the realm of modern biology. Its target is the evidence for design to be found not in the gross anatomical structures of individual species, but within the cell, where recent advances reveal a whole new level of complexity. According to ID, many of the processes now revealed are so complex that they cannot have been assembled by gradual evolution from simpler levels, least of all by a process as haphazard as the natural selection of random mutations.

ID's first big success came in 1991, when the law professor Phillip E. Johnson published his *Darwin on Trial.* Here was a skillfully argued case against Dawkins's thesis that the selection of random mutations could account for complexity and adaptation. Johnson argued that the naturalistic approach to science promoted by atheists like Dawkins arbitrarily ruled out any consideration of the possibility that supernatural processes might be involved. His book was savagely criticized by Stephen Jay Gould and other scientists, allowing the ID movement to float the impression that it was being marginalized by a dogmatic orthodoxy which tolerated no challenges from outsiders. In 1996 the Catholic biochemist Michael J. Behe published what has become the definitive scientific case for ID, his *Darwin's Black Box: The Biochemical Challenge to Evolution.* Behe points to a number of examples in which, he argues, the mechanisms at work within the cell show "irreducible complexity." This is complexity in which every part of the mechanism has to play its role properly or the whole process breaks down. It is thus impossible for such a mechanism to be built up by any step-by-step process. Everything has to come into position at once, and the chance of all this happening simultaneously by random mutations is vanishingly small. Here Dawkins's arguments for the power of selection break down because no intermediate steps are conceivable—Darwin himself had admitted that his theory would fail if it could be shown that this were the case.

The scientific community continues to treat ID with contempt. In some cases, at least, Behe's claims have been undermined by research which has shown that intermediate stages are indeed functional, often because processes are adapted from pre-existing ones that had a very different function. More generally, it is pointed out that there is no active scientific research based on ID. The movement's arguments are always negative: it claims that here is something you will *never* explain—and the whole point of science is to identify a puzzle and to propose naturalistic hypotheses as potential explanations. If the ID movement's argument is accepted in any one case, science simply has to give up at that point, so ID is not so much a form of science as an excuse for stopping science in its tracks.

Nevertheless, the public success of Johnson's and Behe's books has allowed ID to become the platform from which a new generation of efforts to introduce creationism into the schools has been launched. The movement is now well organized, with offices at the Center for the Renewal of Science and Culture, based at the Discovery Institute in Seattle, and a journal, *Origins and Design*. In the new millenium, active campaigns to influence science teaching have been launched in Ohio, Pennsylvania, and Kansas. As this book has been written, the press has reported the resulting hearings, school-board elections, and court cases. Early in 2006 the voters in Dover, Pennsylvania, ejected eight pro-creationist members of their school board following a court case in which the judge decisively upheld a challenge brought by some parents against the board's decision to expose students to ID. In Kansas there is an ongoing battle by fundamentalist religious leaders seeking to have students taught that there are scientifically valid alternatives to evolutionism. The American Association for the Advancement of Science and other similar bodies continue to publish dire warnings about the future of American science if such campaigns undermine their efforts to boost the often deplorable level of science teaching in the nation's schools (see Campbell and Meyer, eds., 2004; Forrest and Gross, 2004; Pennock, 2000, 2001; Scott, 2004; Shanks, 2004).

Despite its central role as the spearhead for the latest efforts to bring anti-evolutionism into the schools, many creationists remain profoundly dissatisfied with ID. At best, it only endorses belief in an abstract Designer for the earliest living cells. It doesn't imply that individual modern species are divinely created, least of all human beings, and its supporters are quite happy with the orthodox scientific model of geological time. Some of them even accept theistic evolution to explain all the later developments in the history of life. For the premillenial Christian groups who depend on the veracity of the Bible to uphold their vision of an imminent Second Coming of Christ, this is not enough—indeed it is totally unacceptable. Young-earth creationism still flourishes in the United States, with some polls suggesting that nearly half the population supports this view. Much of the debate is now conducted on the internet, with websites such as Ken Ham's "Answers in Genesis" (answersingenesis.org) promoting the young-earth position, while others such as "Talk Origins" (talkorigins.org) support a more balanced debate.

What many people really want is creation science, i.e., flood geology, and there are plenty of places where they can find support for this position, even if it cannot be presented in the schools. In addition to many books and websites, there are museums dedicated to showing that dinosaurs walked the earth alongside human beings—creationists believe that they just didn't make it onto the Ark. Hopes that the remains of the Ark will be found remain high, with many creationists seeing the biblical story as a perfectly plausible account of how all life on earth can be traced back to those creatures that survived the flood. They seldom address the questions that would occur to a biogeographer. How did the different species disperse from Mount Ararat to their present locations? One can take a tour of the Grand Canyon with a guide who will explain how all the strata were laid down neatly beneath the flood, the canyon itself being gouged out by the retreating waters. There are no equivalent tours to places where the strata are visibly twisted, intruded with volcanic lava, or deposited unconformably upon one another. The evidence that forced geologists to abandon the flood theory over two centuries ago is simply ignored.

The scientific community remains horrified at the ignorance of basic information that sustains belief in such an outdated theory of the earth. But in a sense it isn't the ignorance that is the problem—it is the level of positive commitment among fundamentalist Christians to a vision of history based on a literal reading of the Bible. There is a sense of insecurity abroad in the world paralleling if not exceeding that which drove the first wave of American fundamentalism in the early twentieth century. People feel their whole world is crumbling around them, despite or even because of the rapid pace of technological progress. In these circumstances people need the kind of reassurance that can be given only by total commitment to a belief system which promises salvation and clearly identifies the supernatural source of that salvation. This level of commitment requires adherence to a worldview defined by a single text, and the chosen text then acquires a degree of authority that prompts a literalist reading even where such a reading defies the body of expert opinion. In an age facing overwhelming political and environmental challenges, many are seeking this level of reassurance. But as opponents such as Dawkins and Dennett are keen to point out, this kind of religious belief can all too easily lead to dogmatism, intransigence, and intolerance. If what I believe is absolutely true, then rival beliefs offered by others must be false and their influence should be curtailed.

What is perhaps most worrying to scientists is the spread of fundamentalism outside the realm of Protestant America (Numbers, 2006). Creationism is flourishing in Mexico and in other countries where the Roman Catholic Church is dominant, including eastern Europe. In August 2005 Cardinal Christoph Schönborn of Vienna hit the headlines with an article in the *New York Times* seeking to promote a creationist position within the Catholic Church. The Church had seemed to accept evolutionism, at least as far as the origin of the human body is concerned, as acknowledged by Pope John Paul II in 1996 (O'Leary, 2006). But even this limited level of accommodation with science is unsatisfactory to some traditionalists, and many suspect that the new Pope, Benedict XVI, will take a much harder line than his predecessor on this issue. Schönborn's attack on evolutionism was vigorously rebutted at the time by Father George Coyne of the Vatican Observatory, but it is significant that Coyne no longer occupies his position as the chief scientific advisor to the Pope.

Outside North America, evangelical Protestantism is enjoying a revival, and spreading its influence into countries once dominated by the Catholic Church. In Europe this movement is often led by churches from immigrant groups that are, in effect, seeking to re-Christianize the continent following its long slide into secularism. Even among the elite, fundamentalist views are spreading, and with them comes opposition to evolutionism. In 2002 Britain was rocked by a controversy sparked by the teaching of creationism in the faith-based Emmanuel City Technology College at Gateshead. Here the influence of a local businessman who sponsors the school generates real concerns about the ability of those with money to direct the education of young people in their community. The school's decision to teach creationism was defended by the prime minister, Tony Blair, and criticized by the archbishop of Canterbury.

Beyond the Western world, Islamic fundamentalism has also taken root, and it too promotes opposition to evolutionism because the theory contradicts traditional beliefs based on the Koran. The West bewails the influence of radical Islam because it sees it as the source of the terrorism now identified as the greatest threat to its culture and its power. It is worth remembering, though, that the sense of insecurity and alienation that drives Muslim extremists may have some parallels with the motivations that have generated the rise of Christian fundamentalism in America and elsewhere. It is the resulting dogmatic reliance on a single text that drives the opposition to evolutionism. In this sense, any fundamentalism will be likely to reject the teachings of science on a contentious issue such as human origins. Dawkins and the militant secularists also point out that it is the existence of rival "sacred" texts which drives much of the conflict racking the world beyond the confines of the scientific community.

THE MIDDLE WAY

This outline of the modern conflicts might generate the feeling that T. H. Huxley, J. W. Draper, and the old advocates of a war between science and religion were correct after all. But throughout this survey of the interaction between religious thought and evolutionism we have seen that there were both religious scientists and liberal theologians who sought a compromise in which the essential aspects of faith could be retained along with a scientific worldview. That compromise has clearly taken serious damage in the course of the twentieth century and might seem now to be in a state of terminal decline. Has the middle way been effectively eliminated, leaving us with only the alternatives of atheistical Darwinism and religious fundamentalism? This is the view promoted by extremists on both sides, including Dawkins and Dennett for the atheists and a host of evangelical preachers for the creationists. But clearly there is a middle ground, at least at the level of intellectual debate. There are still many who actively promote a synthesis between science and religion, and are determined to include evolutionism in the package. Our survey of the modern debates will conclude with these efforts at continued reconciliation. Indeed, we shall see that there have been notable developments since the emergence of modern Darwinism which may put religious believers in a better position than they have ever been before to welcome the selection theory and its implications for humanity. And as Michael Ruse has argued against the militant atheism of Dennett, in a world where many are strongly attracted to some form of religious belief, it is counterproductive for the evolutionist to link the theory too rigidly to materialism when other interpretations are possible.

Perhaps this survey of the middle way and its antecedents will help the cause of those who want to resist the growing influence of fundamentalism on both sides. But it has to be conceded that the possibility of turning the fundamentalist tide in religion seems remote. If the true foundation of literalism is a psychological need to believe, no amount of intellectual argument will succeed in undermining such a faith. The danger is that the liberal theologians and intellectuals debate in their ivory tower while militant religion runs rampant in the everyday world. But there may still be some hope. The fundamentalists on both sides bolster their arguments with the claim that there is no middle ground. There is only a black-and-white alternative, and you must choose one side or the other. Undermining the credibility of this artificial choice may not play much of a role in the wider world, but at least it's worth a try.

In the decades following World War II many liberal Christians were attracted to what turned out to be the last gasp of the old non-Darwinian vision of evolution based on inevitable progress. The best-known exponent of this approach was a French Jesuit priest,

Pierre Teilhard de Chardin (Lukas and Lukas, 1977). Teilhard was a paleontologist and so could speak with some appearance of authority on evolutionism. He always claimed that his system was purely scientific. In fact it was a compendium of the kind of non-Darwinian thinking still popular in the interwar years with a vision of cosmic spiritual progress. God's influence was at work within the world, pushing evolution in the direction of increased levels of moral and spiritual awareness. Although in some respects reminiscent of Bergson's creative evolution, Teilhard insisted that the direction of progress was rigidly predetermined. Humans were the inevitable outcome, and our future spiritual progress would eventually lead to an "omega point" at which we would merge into a single spiritual unit and unite with the Creator.

Not surprisingly, the Catholic Church refused to let Teilhard publish during his lifetime. The Church was now prepared to allow discussion of the claim that the human body had been produced by evolution, but any suggestion that the soul had evolved from animal antecedents was anathema. Teilhard's works were eventually published posthumously, the English translation of his *Phenomenon of Man* appearing in 1959. There was a good deal of excitement in liberal religious circles, seeming almost to revive the hopes of a reconciliation with nonmaterialistic science that had been popular among the Modernists of the 1920s. These hopes were boosted when Julian Huxley (always an enthusiast for progress) penned an introduction to the translation of *The Phenomenon of Man*. Charles Raven, whom we last met as an anti-Darwinian enthusiast for creative evolution before the war, now endorsed Teilhard's vision as identical to his own, and lamented that they had never met (Raven, 1962). Raven's Gifford lectures of 1951 were also a belated effort to revive this older vision of evolution in the hopes of reconciling it with liberal, postmillennial Christianity (Raven, 1953).

Unfortunately, the scientists were no longer respecting their end of the bargain. Most were appalled that Huxley could be misled by his optimistic ideology into endorsing a teleological evolutionism so obviously out of tune with the new Darwinism which he himself had helped to create. More typical of the scientists' reaction was a vitriolic review of Teilhard's book by the British biologist Peter Medawar, which exposed it as empty rhetoric with nothing to offer on the actual causes of evolution (reprinted in Medawar, 1982: 242–251). In the new environment created by the triumph of modern synthetic Darwinism within biology, liberal religious thinkers could no longer get away with appeals to outdated evolution theories that had been designed to sidestep the materialistic implications of Darwinism. If religion was hoping to do a deal with science, it had to face up—at long last—to the challenge of an evolutionary mechanism based on the natural selection of randomly generated variations.

The challenges remain the same as those identified by Wilberforce and the other early critics of the selection theory. Can a Christian really believe that humans are the products of a trial-and-error process driven by struggle and death? How can this be reconciled with the belief that the world was designed by God, or that we have immortal souls which will face a final judgment? A liberal Christian may be able to accept that we are the products of a natural process, but surely that process must show some evidence that the Creator who started it all going intended it to have its outcome in something very like humanity. The complete separation of human from animal nature will have to be abandoned along with a literal interpretation of Genesis. That might be possible, but what then of the Christian view that we are a fallen race separated from God? The philosopher Michael Ruse, who is not himself a Christian, has argued (2001, 2003) that Darwinism is compatible with a meaningful notion of design, and that a liberal Christian should—with some difficulty, perhaps—be able to live with the model of human nature established by modern Darwinism. Whether or not anything resembling the idea of Original Sin can be preserved is a complex question, as is the responsibility of God for allowing His design to be worked out by a process based on suffering and death.

How might a liberal Christian go about reconciling what have so long been seen as two mutually incompatible visions of the universe and the human situation? One popular move against the materialism of modern science does not seem very promising. This is the so-called "process theology" derived from the philosophy of Alfred North Whitehead. Here the basic elements of reality are seen as events, not objects, and God appears as the background of experience and the potential for future development. Apart from being totally incomprehensible to most people (the present writer included), this approach simply ignores the issues posed by biology. It might be able to deal with the complex issues raised by modern physics' reformulation of our ideas about the ultimate nature of reality, but it is too fine-grained to offer much help with ideas about genes and natural selection.

Where theologians' efforts to address the nature of matter do offer something of relevance is in the area of what is known as the cosmological anthropic principle (Barrow and Tippler, 1986). As Arthur Peacocke (1986) and John Polkinghorne (1994, 1998a, 1998b) show, here is a reformulation of the design argument which—unlike ID in biology—really has forced scientists to think about whether or not the ultimate nature of the universe was "fine-tuned" from the beginning in a way that made it possible, perhaps even inevitable, for complex structures such as living things to emerge. If the various physical constants that define the nature of matter were even slightly different from what they actually are, stars and planets could not form, and life could never evolve. There is no room here to go behind the scenes of these bold assertions, but the anthropic principle is important to our argument because it does depend on the idea of evolution. To unpack the potentialities that theologians see in the basic nature of the universe, there has to be a process by which life can emerge on suitable planets and then evolve toward more complex forms. In a very remote sense, the anthropic principle seems to be telling us that the universe was set up in such a way that complex living things were bound to appear sooner or later.

The problem for the Christian is that there is very little sense here that morally responsible, self-aware creatures like ourselves are inevitable. The anthropic principle requires evolutionary progress, but does not specify the outcome on any particular planet where life might appear. And given that the hard-line Darwinists like Dawkins insist that progress is not inevitable on a world driven by natural selection, can we even be sure that anything more interesting than an amoeba will ever evolve anywhere? For this reason, some theologians sympathetic to evolutionism still feel that a more purposeful process than natural selection must be involved. But this approach runs the risk of reintroducing the old non-Darwinian ideas that science has now discredited.

A more promising approach might be to challenge the ultra-Darwinians' insistence on the completely nonprogressive character of natural selection. Even Dawkins allows for a kind of progress, as when predator and prey get involved in an "arms race" in which each drives the other to run faster and faster. Such improvements relate to a specific ecological relationship, of course, but evolutionary progressionism has always tended to assume that—in the long run, at least—there will be a tendency for those species with the best "on-board computers" (i.e., brains) to succeed over those less well endowed. Insisting that such a faith in progress is anthropocentric and teleological has become a defining characteristic of modern ultra-Darwinism. But it might be queried whether or not the absolute prohibition on any concession to the idea of progress is not itself a product of the materialist ideology of those who expound this extreme form of selectionism.

This is certainly the position of those religious thinkers who are still looking for a way to reconcile Darwinism with some form of purpose in nature. But there are other developments to which they can also appeal. Writers such as Stuart Kauffman (1995) have argued that even in inorganic nature there are processes in which complexity seems to emerge from fairly simple interactions and then maintain itself as an organized system. Perhaps such a tendency is also applicable in organic evolution. There are also growing doubts about the validity of Dawkins's basic assumption that all evolution depends on the natural selection of small genetic differences. The new science of evolutionary developmental biology (evo-devo) suggests that there are many aspects of organic structure which are controlled by deep-seated developmental mechanisms common to widely different branches of the tree of life (Carroll, 2005). On this model, we don't have to envisage every structure as individually constructed in a series of small steps. On the contrary, living things have an array of genetically founded building blocks they can call upon and adapt to a wide range of different purposes. Once those building-blocks were in place (and they would have to have been formed at a very early stage indeed to be distributed so widely), they could help to manufacture the complex structures that creationists insist are beyond the scope of step-by-step natural selection.

Most Darwinians would insist that to go down this route, theologians must accept that there is no built-in trend toward humanity. Darwinism offers only an open-ended, haphazard, and largely unpredictable model of progress. At best, we can hope that something as complex and self-aware as human beings would eventually emerge in the universe. There are a few biologists who have now begun to argue that evolution might not be as open-ended as the original Darwinists believed. Simon Conway Morris is a leading expert on the famous Burgess shale fossils, which give us an unrivaled glimpse into the diversity of life immediately following the "Cambrian explosion," in which complex animals first appear in the record. We have noted how Stephen Jay Gould (1989) used the bizarre character of these fossils to drive home his point that the history of life on earth could very easily have given rise to a world completely different to the one we live in. But Morris has challenged Gould's interpretation of these fossils and has now gone on to argue that the course of evolution is far more rigidly constrained than we used to think (Morris, 2003). He points to the phenomenon of convergence, in which evolution repeatedly comes up with very similar structures in creatures exposed to the same environment. He believes that this effect is so pervasive that we can, after all, predict fairly accurately what forms of life are possible. Perhaps the human race really is an inevitable outcome of cosmic evolution, even in a Darwinian universe.

Many biologists find Morris's point overstated, and in fact most liberal theologians have grown used to the idea that we cannot expect humans—two legs, two arms, and all—to appear wherever evolution takes off. What the believer needs is some hope that creatures with a moral and spiritual capacity must eventually emerge, whatever their physical form. The main problems then facing them are to explain how an apparently materialistic process can generate these capacities and to reconcile the endless suffering at the heart of nature—and of the evolutionary mechanism—with their conception of God the Creator. Surprisingly, these perennial problems seem to be at last yielding to analysis in a way that allows the Darwinian process to be seen as conformable with a certain kind of Christian faith.

The question of human nature is too broad to be explored in detail here. Modern evolutionary psychology, driven by the logic of sociobiology, presents human nature as a balance between inherited instincts and the freedom to learn and innovate. The instincts form the core of human nature, and if the sociobiological approach is accepted, they inevitably set up a tension between selfish and altruistic drives. The fact that we have to juggle these conflicting tendencies can be seen as the basis for a Christian view of human nature. Just as the early-twentieth-century Modernists argued, the legacy of our animal past represents a core of selfishness that can form the basis for a view of human nature as "flawed" in the sense that it constantly tempts us to do things which our altruistic instincts and our cultural conditioning regard as evil. The important thing is that we do have some degree of freedom to negotiate our response to these conflicting impulses. Our values are not solely determined by our biology (Rolston, 1999).

It is this element of freedom that seems to offer theologians their best chance of understanding why God would have chosen an undirected process like natural selection to serve as the driving force of His creation. In the end, the universe is built in such a way that evolution will eventually generate morally aware beings like ourselves (Ward, 1996). But if evolution were driven by a rigidly predetermined trend toward an inevitable goal, it is difficult to see how the creatures produced by such a trend could have any freedom of action. In the eyes of theologians such as Polkinghorne, Peacocke, and Ward, and of religious scientists such as Kenneth R. Miller (1999), natural selection, for all its haphazardness and unpredictability, may be the only kind of mechanism that can do the job that the Creator intended. They suggest that in order to give His creation the freedom to develop its own independent character, He did not lay down a fixed goal for it to achieve. On the contrary, he emptied Himself into the universe and allowed it the freedom to develop in its own way, as directed by the creatures that evolved within it. He voluntarily took the risk that things might not turn out perfectly in order to give us the gift of freedom and responsibility.

But what about the pain and suffering that are so prominent a feature of nature, and which seem essential for evolution to occur? It is worth noting that to many fundamentalist Christians, suffering and death only came into the world as a consequence of human sin. Their rejection of Darwinism is coupled with a refusal to accept that there was a struggle for existence in the animal kingdom before Adam disobeyed his Creator. Such a position is unacceptable to the scientists who see the evidence for predation throughout the fossil record. To accept that suffering is a central feature of the world presents a problem for religion in general, but it may offer an opportunity for liberal Christians who are prepared to think more flexibly about the relationship between God and humanity, as manifested in the life and death of Christ.

Here the thought of John Polkinghorne and John F. Haught (2000, 2004) becomes instructive, because they see that the central role played by suffering in the world may be just what we should expect if God had relinquished His control over nature in order to give His creatures a degree of freedom within their world. Unlike some other religions, Christianity can be presented as a religion in which God, far from sitting outside His creation, has actually entered into it and suffers along with the struggling creatures within it. Such a vision seems to make sense of the fact that the son of God himself suffered the consequences of human selfishness and intolerance—and the Father did not intervene to prevent this supreme level of involvement and sacrifice. As Polkinghorne writes:

> In the lonely figure hanging in the darkness and dereliction of Calvary the Christian believes that he sees God opening his arms to embrace the bitterness of the strange world he has made. The God revealed in the vulnerability of the incarnation and the vulnerability of creation are one. He is the crucified God, whose paradoxical power is perfected in weakness, whose self chosen symbol is the King reigning from the gallows (Polkinghorne, 1989:68).

Powerful stuff, even for a nonbeliever like myself. Here is a totally different vision of the relationship between God, humanity, and nature to that offered by the fundamentalists. This is not a God who punishes us eternally unless we accept His son's sacrifice as the only route back into His favor. It is a God who participates in the human drama and in the drama of creation, and if there is any kind of God who makes sense to the convinced Darwinian, this is probably it.

No fundamentalist will accept such a rival vision of the Christian message, and there is little chance that evolutionists will benefit in the short term from any mass movement toward Polkinghorne's position among American Christians. But the fact that liberal Christian thinkers can now articulate a vision that seems almost to welcome those aspects of Darwinism long regarded as incompatible with any form of religious faith shows that the renewed state of war between fundamentalists and atheistic Darwinists is not the only game in town. This is certainly a postmillenial vision which hopes that we can achieve something worthwhile in this world without external divine intervention. But because it sees God as struggling in the world alongside us, it is a long way from the old progressionist approach once favored by liberals. There is no design planned out in advance for the world. We have to work it out for ourselves, and there is no guarantee we are going to succeed—that is the chance that God has taken.

The history of the relationship between science and religion shows that there have always been religious thinkers looking for a middle way that will allow them to accept the latest developments of science. What the last chapter of this story has revealed is that new developments have been possible within this tradition, developments that have made possible an entirely new kind of natural theology. We no longer expect an external designer who has imposed order on the world from without. And we can see our own imperfect natures as products of the process by which He has chosen to allow the world to develop. At a time when extremists on both sides want to convince us that we must be either for or against their own dogma, anything that offers the prospect of dialogue is worthwhile. Even those of us who can't accept Polkinghorne's God may be able to sympathize with the humanizing trend within religion that it represents. The case for a rational approach to the study of origins can only benefit from such a trend.

BIBLIOGRAPHY

INDEX

Adams, Mark B., ed. 1994. *The Evolution of Theodosius Dobzhansky*. Princeton: Princeton University Press.

Anon, 1925. *The World's Most Famous Trial: Tennessee Evolution Case*. Cincinnati, OH: National Books Co.

Appel, Toby A. 1987. *The Cuvier-Geoffroy Debate: French Biology in the Decades before Darwin*. Oxford: Oxford University Press.

Appleby, R. Scott, 2001. "Exposing Darwin's 'Hidden Agenda': Roman Catholic Responses to Evolution, 1875–1925." In Ronald L. Numbers and John Stenhouse, eds., *Disseminating Darwinism: The Role of Place, Race, Religion and Gender*, 173–207. Cambridge: Cambridge University Press.

Appleman, Philip, ed. 2001. *Darwin: A Norton Critical Edition*. 3rd ed. New York: Norton.

Bannister, Robert C. 1979. *Social Darwinism: Science and Myth in Anglo-American Social Thought*. Philadelphia: Temple University Press.

Barbour, Ian G. 1966. *Issues in Science and Religion*. Englewood Cliffs, NJ: Prentice Hall.

———. 1968. *Science and Religion: New Perspectives on the Dialogue*. New York: Harper and Row.

Barnes, Ernest William, 1927. *Should Such a Faith Offend? Sermons and Addresses*. London: Hodder and Stoughton.

———. 1933. *Scientific Theory and Religion: The World Observed by Science and its Spiritual Interpretation*. Cambridge: Cambridge University Press.

Barnes, John. 1979. *Ahead of his Age: Bishop Barnes of Birmingham*. London: Collins.

Barr, Alan P., ed. 1997. *Thomas Henry Huxley's Place in Science and Letters*. Athens, GA: University of Georgia Press.

Barrow, J. D., and F. J. Tippler. 1986. *The Anthropic Cosmological Principle*. Oxford: Oxford University Press.

Barthélemy-Madaule, Madeleine. 1982. *Lamarck the Mythical Precursor: A*

Study of the Relations between Science and Ideology. Cambridge, MA: MIT Press.

Bartholomew, Michael. 1973. "Lyell and Evolution: An Account of Lyell's Response to the Prospect of an Evolutionary Ancestry for Man." *British Journal for the History of Science,* 6: 261–303.

———. 1975. "Huxley's Defence of Darwinism." *Annals of Science* 32: 525–535.

Bateson, William. 1894. *Materials for the Study of Variation: Treated with Especial Regard to Discontinuity in the Origin of Species.* London. Reprint with foreword by Peter J. Bowler and introduction by Gerry Webster. Baltimore: Johns Hopkins University Press, 1992.

Beecher, Henry Ward. 1885. *Evolution and Religion.* London: James Clark.

Behe, Michael. 1996. *Nature's Black Box: The Biochemical Challenge to Evolution.* New York: Simon and Schuster.

Bergson, Henri. 1911. *Creative Evolution.* Trans. Arthur Mitchell. New York: Henry Holt.

Blinderman, Charles. 1986. *The Piltdown Inquest.* Buffalo, NY: Prometheus Books.

Bowler, Peter J. 1973. "Bonnet and Buffon: Theories of Generation and the Problem of Species." *Journal of the History of Biology* 6: 259–281.

———. 1976a. *Fossils and Progress: Paleontology and the Idea of Progressive Evolution in the Nineteenth Century.* New York: Science History Publications.

———. 1976b. "Malthus, Darwin and the Concept of Struggle." *Journal of the History of Ideas* 37: 631–650.

———. 1983. *The Eclipse of Darwinism: Anti-Darwinian Evolution Theories in the Decades around 1900.* Baltimore: Johns Hopkins University Press.

———. 1986. *Theories of Human Evolution: A Century of Debate, 1844–1944.* Baltimore: Johns Hopkins University Press. Oxford: Basil Blackwell.

———. 1988. *The Non-Darwinian Revolution: Reinterpreting a Historical Myth.* Baltimore: Johns Hopkins University Press.

———. 1989. *The Mendelian Revolution: The Emergence of Hereditarian Concepts in Modern Science and Society.* Baltimore: Johns Hopkins University Press.

———. 1990. *Charles Darwin: The Man and His Influence.* Oxford: Basil Blackwell, reprinted Cambridge: Cambridge University Press.

———. 1996. *Life's Splendid Drama: Evolutionary Biology and the Reconstruction of Life's Ancestry, 1860–1940.* Chicago: University of Chicago Press.

———. 1998. "Evolution and the Eucharist: Bishop E. W. Barnes on Science and Religion in the 1920s and 1930s." *British Journal of the History of Science,* 31: 453–467.

———. 2001. *Reconciling Science and Religion: The Debates in Early Twentieth-Century Britain.* Chicago: University of Chicago Press.

———. 2003. *Evolution: The History of an Idea.* 3rd ed. Berkeley: University of California Press.

———. 2005a. "From Science to the Popularization of Science: The Career of J. Arthur Thomson." In David Knight and Matthew D. Eddy, eds., *Science and Beliefs: From Natural Philosophy to Natural Science,* 231–250. Aldershot: Ashgate.

———. 2005b. "The Specter of Darwinism: The Popular Image of Darwinism in Early Twentieth-Century Britain." In Abigail Lustig, Robert J. Richards, and Michael Ruse, eds., *Darwinian Heresies,* 48–68. New York: Cambridge University Press.

Brooke, John Hedley. 1991. *Science and Religion: Some Historical Perspectives.* Cambridge: Cambridge University Press.

———. 2001. "The Wilberforce-Huxley Debate—Why Did It Happen?" *Science and Christian Belief,* 13: 127–141.

Browne, Janet. 1989. "Botany for Gentlemen: Erasmus Darwin and *The Loves of the Plants.*" *Isis* 80: 593–621.

———. 1995. *Charles Darwin: Voyaging.* London: Jonathan Cape.

———. 2002. *Charles Darwin: The Power of Place.* London: Jonathan Cape.

Brundell, Barry. 2001. "Catholic Church Politics and Evolution Theory, 1894–1902." *British Journal of the History of Science* 34: 81–96.

Buckland, William. 1823. *Reliquiae Diluvianae: Or Observations of the Organic Remains contained in Caves, Fissures and Diluvial Gravel, and other Geological Phenomena, attesting the Action of a Universal Deluge.* London: reprinted New York: Arno Press, 1977.

———. 1836. *Geology and Mineralogy Considered with Reference to Natural Theology.* 2 vols. London.

Buffon, Georges Louis Leclerc, Comte de. 1981. *From Natural History to the History of Nature: Readings from Buffon and His Critics.* Ed. Philip R. Sloan and J. Lyon. Notre Dame: University of Notre Dame Press.

Burchfield, Joe D. 1975. *Lord Kelvin and the Age of the Earth.* New York: Science History Publications.

Burkhardt, Richard W., Jr. 1977. *The Spirit of System: Lamarck and Evolutionary Biology.* Cambridge, MA: Harvard University Press. Rev. ed. 1995.

Burnet, Thomas. 1691. *The Sacred Theory of the Earth.* London: reprinted with an introduction by Basil Willey, London: Centaur Press, 1965.

Butler, Samuel. 1879. *Evolution, Old and New: Or the theories of Buffon, Dr. Erasmus Darwin, and Lamarck, as Compared with that of Mr. Charles Darwin.* London: Hardwicke and Bogue.

———. 1908. *Essays on Life, Art and Science.* London: reprinted Port Washington, NY: Kennikat Press, 1970.

Campbell, John Angus, and Stephen A. Meyer, eds., 2004. *Darwinism, Design and Public Education.* East Lansing, MI: Michigan State University Press.

Campbell, R. J. 1907. *The New Theology.* London: Chapman and Hall.

Caplan, Arthur L., ed. 1978. *The Sociobiology Debate.* New York: Harper & Row.

Carpenter, Joel A., ed. 1988. *Fundamentalism in American Religion, 1880–1950.* 45 vols. New York: Garland.

Carroll, Sean B. 2005. *Endless Forms Most Beautiful: The New Science of Evo Devo and the Making of the Animal Kingdom.* London: Weidenfeld and Nicolson.

Chambers, Robert. 1994. *Vestiges of the Natural History of Creation and Other Evolutionary Writings.* Ed. James Secord. Chicago: University of Chicago Press.

Clark, Constance A. 2001. "Evolution for John Doe: Pictures, the Public, and the Scopes Trial Debate." *Journal of American History,* 87: 1275–1303.

Clements, Keith W. 1988. *Lovers of Discord: Twentieth-Century Theological Controversies in England.* London: SPCK.

Coleman, William. 1964. *Georges Cuvier, Zoologist: A Study in the History of Evolution Theory.* Cambridge, MA: Harvard University Press.

Cooter, Roger. 1985. *The Cultural Meaning of Popular Science: Phrenology and the Organization of Consent in Nineteenth-Century Britain.* Cambridge: Cambridge University Press.

Cope, Edward Drinker. 1868. "The Origin of Genera." *Proceedings of the Academy of Natural Sciences, Philadelphia,* 20: 242–300.

———. 1887a. *The Origin of the Fittest: Essays in Evolution.* Reprinted with Cope, *The Primary Factors of Organic Evolution.* New York: AMS Press, 1974.

———. 1887b. *Theology of Evolution: A Lecture.* Philadelphia: Arnold.

Corsi, Pietro. 1988a. *The Age of Lamarck: Evolutionary Theories in France, 1790–1830.* Berkeley: University of California Press.

―――. 1988b. *Science and Religion: Baden Powell and the Anglican Debate, 1800–1860.* Cambridge: Cambridge University Press.

Cuvier, Georges. 1817. *An Essay on the Theory of the Earth.* Translated by Robert Kerr, with notes by Robert Jameson. 3rd ed. Edinburgh: reprinted New York: Arno Press, 1977.

Daniels, George. 1968. *Darwinism Comes to America.* Waltham, MA: Blaisdell Publishing Co.

Däniken, Erich von. 1970. *Chariots of the Gods? Unsolved Mysteries of the Past.* Trans. Michael Heron. New York: C. P. Putnam's Sons.

―――. 1977. *According to the Evidence: My Proof of Man's Extraterrestrial Origins.* Trans. Michael Heron. London: Souvenir Press.

Darwin, Charles Robert. 1859. *On the Origin of Species by Means of Natural Selection: Or the Preservation of Favoured Races in the Struggle for Life.* London: reprinted with an introduction by Ernst Mayr. Cambridge, MA: Harvard University Press, 1964. (Subsequent editions of Darwin's book were simply titled *Origin of Species.*)

―――. 1871. *The Descent of Man and Selection in Relation to Sex.* 2 vols. London: John Murray. 1895 ed. reprinted New York: AMS Press, 1972.

―――. 1975. *Charles Darwin's Natural Selection: Being the Second Part of His Big Species Book Written from 1856 to 1858.* Ed. Robert C. Stauffer. London: Cambridge University Press.

―――. 1987. *Charles Darwin's Notebooks (1836–1844).* Ed. Paul H. Barrett et al. Cambridge: Cambridge University Press.

Darwin, Charles, and Alfred Russel Wallace. 1958. *Evolution by Natural Selection.* With a foreword by Sir Gavin de Beer. Cambridge: Cambridge University Press.

Darwin, Erasmus. 1794–96. *Zoonomia: Or the Laws of Organic Life.* 2 vols. London: reprinted New York: AMS Press, 1974.

―――. 1803. *The Temple of Nature.* London: reprinted Elmsford, NY: Pergamon (British Book Center).

Davis, Edward B. 2005. "Science and Religious Fundamentalism in the 1920s." *American Scientist,* 93: 253–260.

Dawkins, R. 1976. *The Selfish Gene.* Oxford: Oxford University Press.

―――. 1982. *The Extended Phenotype: The Gene as the Unit of Selection.* San Francisco: W. H. Freeman.

―――. 1986. *The Blind Watchmaker.* Harlow: Longman Scientific and Technical.

―――. 1995. *River out of Eden: A Darwinian View of Life.* London: Phoenix.

———. 1996. *Climbing Mount Improbable*. New York: Norton.

———. 2006. *The God Delusion*. Boston: Houghton Mifflin.

Dawson, J. W. 1890. *Modern Ideas of Evolution*. Reprinted ed. William R. Shea and John F. Cornell. New York: Prodist.

De Bont, Raf. 2005. "Rome and Theistic Evolutionism: The Hidden Strategies behind the 'Dorlodot Affair', 1920–1926." *Annals of Science*, 62: 457–478.

De Camp, L. S. 1968. *The Great Monkey Trial*. Garden City, NY: Doubleday.

De Grazia, A. 1966. *The Velikovsky Affair*. New Hyde Park, NY: University Books.

Dennett, Daniel. 1995. *Darwin's Dangerous Idea: Evolution and the Meaning of Life*. New York: Simon and Schuster.

———. 2006. *Breaking the Spell: Religion as a Natural Phenomemon*. New York: Viking.

Desmond, Adrian. 1982. *Archetypes and Ancestors: Palaeontology in Victorian London, 1850–1875*. London: Blond and Briggs; reprinted Chicago: University of Chicago Press.

———. 1989. *The Politics of Evolution: Morphology, Medicine and Reform in Radical London*. Chicago: University of Chicago Press.

———. 1994. *Huxley: The Devil's Disciple*. London: Michael Joseph.

———. 1997. *Huxley: Evolution's High Priest*. London: Michael Joseph.

Desmond, Adrian, and James R. Moore. 1991. *Darwin*. London: Michael Joseph.

Diderot, Denis. 1966. *D'Alembert's Dream and Rameau's Nephew*. Trans. L. W. Tancock. Harmondsworth, Middlesex: Penguin Books.

Di Gregorio, Mario A. 1984. *T. H. Huxley's Place in Natural Science*. New Haven: Yale University Press.

Dilliston, F. W. 1975. *Charles Raven: Naturalist, Historian, Theologian*. London: Hodder and Stoughton.

Dobzhansky, Theodosius. 1962. *Mankind Evolving*. New Haven: Yale University Press.

———. 1967. *The Biology of Ultimate Concern*. New York: New American Library.

Dorlodot, Henri de. 1925. *Darwinism and Catholic Thought*. Trans. E. Messenger. New York: Benziger Brothers.

Draper, J. W. 1875. *History of the Conflict between Religion and Science*. London.

Drummond, Henry. 1894. *The Ascent of Man*. New York: James Pott.

Dupree, A. Hunter. 1959. *Asa Gray*. Cambridge, MA: Harvard University Press. Reprinted Baltimore: Johns Hopkins University Press, 1988.

Durant, John R., ed. 1985. *Darwinism and Divinity: Essays on Evolution and Religious Belief.* Oxford: Basil Blackwell.

Elder, Gregory P. 1996. *Chronic Vigor: Darwin, Anglicans, Catholics and the Development of a Doctrine of Providential Evolution.* Lanham, MD: University Press of America.

Ellegård, Alvar. 1958. *Darwin and the General Reader: The Reception of Darwin's Theory of Evolution in the British Periodical Press, 1859–1872.* Goteburg: Acta Universitatis Gothenburgensis. Reprinted Chicago: University of Chicago Press, 1990.

Ferngren, Gary B., ed. 2002. *Science and Religion: A Historical Introduction.* Baltimore: John Hopkins University Press.

Fichman, Martin. 1984. "Ideological Factors in the Dissemination of Darwinism in England, 1860–1900." In E. Mendelsohn, ed., *Transformation and Tradition in the Sciences: Essays in Honor of I. Bernard Cohen*, 471–485. Cambridge: Cambridge University Press.

———. 2004. *An Elusive Victorian: The Evolution of Alfred Russel Wallace.* Chicago: University of Chicago Press.

Fisher, R. A. 1930. *The Genetical Theory of Natural Selection.* Oxford: Clarendon Press, reprinted New York: Dover, 1958.

———. 1950. *Creative Aspects of Natural Law.* Cambridge: Cambridge University Press.

Fiske, John. 1874. *Outlines of Cosmic Philosophy.* Boston: reprinted with an introduction by David W. Noble, New York: Johnson Reprint Corporation, 1969.

Forrest, Barbara, and Paul R. Gross. 2004. *Creationism's Trojan Horse: The Wedge of Intelligent Design.* New York: Oxford University Press.

Fosdick, Harry Emerson. 1922. *Christianity and Progress.* London: Nisbett.

———. 1926. *Adventures in Religion and Other Essays.* London: SCM.

———. 1932. *As I See Religion.* London: SCM.

———. 1956. *The Living of These Days: An Autobiography.* New York: Harper/London: SCM.

Fyfe, Aileen. 2004. *Science and Salvation: Evangelical Popular Science Publishing in Victorian Britain.* Chicago: University of Chicago Press.

Gayon, Jean. 1998. *Darwinism's Struggle for Survival: Heredity and the Hypothesis of Natural Selection.* Cambridge: Cambridge University Press.

Gilbert, James. 1997. *Redeeming Culture: American Religion in an Age of Science.* Chicago: University of Chicago Press.

Gillispie, Charles Coulston. 1959. *Genesis and Geology: A Study in the Relations of Scientific Thought, Natural Theology and Social Opinions in Great Britain, 1790–1859*. Reprinted New York: Harper, 1959.

Ginger, Ray. 1958. *Six Days or Forever.* Boston: Beacon Press.

Gish, Duane T. 1972. *Evolution: The Fossils Say No!* San Diego: Creation Life Publishers.

Glass, Bentley, Owsei Temkin, and William Strauss Jr., eds. 1959. *Forerunners of Darwin, 1745–1859.* Baltimore: Johns Hopkins University Press.

Glick, Thomas F., ed. 1974. *The Comparative Reception of Darwinism.* Austin: University of Texas Press. New ed. Chicago: University of Chicago Press, 1988.

Goldsmith, Donald, ed. 1977. *Scientists Confront Velikovsky.* Ithaca, NY: Cornell University Press.

Gould, Stephen Jay. 1977. *Ontogeny and Phylogeny.* Cambridge, MA: Harvard University Press.

———. 1981. *The Mismeasure of Man.* New York: Norton.

———. 1987. *Time's Arrow, Time's Cycle: Myth and Metaphor in the Discovery of Geological Time.* Cambridge, MA: Harvard University Press.

———. 1989. *Wonderful Life: The Burgess Shale and the Nature of History.* London: Hutchinson Radius.

———. 1999. *Rocks of Ages: Science and Religion in the Fullness of Life.* New York: Ballantine.

Gray, Asa. 1876. *Darwiniana: Essays and Reviews Pertaining to Darwinism.* New York. Reprint ed. A. Hunter Dupree. Cambridge, MA: Harvard University Press, 1963.

Grayson, Donald K. 1983. *The Establishment of Human Antiquity.* New York: Academic Press.

Greene, John C. 1959. *The Death of Adam: Evolution and Its Impact on Western Thought.* Ames: Iowa State University Press.

———. 1990. "The Interaction of Science and World View in Sir Julian Huxley's Evolutionary Biology." *Journal of the History of Biology* 23: 39–55. Reprinted in Greene 1999, 71–90.

———. 1999. *Debating Darwin: Adventures of a Scholar.* Claremont, CA: Regina Books.

Gruber, Jacob W. 1960. *A Conscience in Conflict: The Life of St. George Jackson Mivart.* New York: Columbia University Press.

Haber, Francis C. 1959. *The Age of the World: Moses to Darwin.* Baltimore: Johns Hopkins University Press.

Haeckel, Ernst. 1876. *The History of Creation: Or the Development of the Earth and Its Inhabitants by the Action of Natural Causes. A Popular Exposition of the Doctrine of Evolution in General and of that of Darwin, Goethe and Lamarck in Particular.* 2 vols. New York: Appleton.

———. 1879. *The Evolution of Man: A Popular Exposition of the Principal Points of Human Ontogeny and Phylogeny.* 2 vols. New York: Appleton.

Haldane, J. B. S. 1932. *The Causes of Evolution.* London: Longmans, Green.

Haller, John S. 1975. *Outcasts from Evolution: Scientific Attitudes of Racial Inferiority, 1859–1900.* Urbana: University of Illinois Press.

Harrison, Peter. 1998. *The Bible, Protestantism, and the Rise of Natural Science.* Cambridge: Cambridge University Press.

Haught, John F. 2000. *God after Darwin.* New York: Westview.

———. 2004. *Darwin, Design, and the Promise of Nature.* London: St Mary-le-Bow.

Hawkins, Mike. 1997. *Social Darwinism in European and American Thought, 1860–1945: Nature as Model, Nature as Threat.* Cambridge: Cambridge University Press.

Helfand, M. S. 1977. "T. H. Huxley's 'Evolution and Ethics': The Politics of Evolution and the Evolution of Politics." *Victorian Studies* 20: 159–177.

Herbert, Sandra. 2005. *Charles Darwin, Geologist.* Ithaca, NY: Cornell University Press.

Himmelfarb, Gertrude. 1959. *Darwin and the Darwinian Revolution.* Reprinted New York: Norton.

Hodge, Charles. 1994. *What Is Darwinism?,* ed. Mark A. Noll and David Livingstone. Grand Rapids, MI: Baker.

Hodge, M. J. S. 1971. "Lamarck's Science of Living Bodies." *British Journal of the History of Science,* 5: 323–352.

———. 1972. "The Universal Gestation of Nature: Chambers' *Vestiges* and *Explanations.*" *Journal of the History of Biology,* 5: 127–152.

———. 1985. "Darwin as a Lifelong Generation Theorist." In Kohn 1985, 207–243.

———. 2005. "Against 'Revolution' and 'Evolution.'" *Journal of the History of Biology,* 38: 101–121.

Hodge, M. J. S., and D. Kohn. 1985. "The Immediate Origins of Natural Selection." In Kohn 1985, 185–206.

Hofstadter, Richard. 1959. *Social Darwinism in American Thought.* Rev. ed. New York: George Braziller.

Hull, David L. 1973. *Darwin and His Critics: The Reception of Darwin's Theory of Evolution by the Scientific Community.* Cambridge, MA: Harvard University Press.

Huxley, Julian S. 1942. *Evolution: The Modern Synthesis.* London: Allen and Unwin.

————, ed. 1961. *The Humanist Frame.* London: Allen and Unwin.

————. 1964. *Essays of a Humanist.* London: Chatto and Windus.

————. 1970. *Memories.* London: Allen and Unwin.

Huxley, T. H. 1863. *Evidence as to Man's Place in Nature.* London: Williams and Norgate. Reprinted in Huxley, *Man's Place in Nature. Collected Essays,* vol. 7. London: Macmillan, 1894.

————. 1894. *Evolution and Ethics. Collected Essays,* vol. 9. London: Macmillan.

Irvine, William. 1955. *Apes, Angels and Victorians: The Story of Darwin, Huxley and Evolution.* London: reprinted Cleveland: Meridian Books, 1959.

Israel, Jonathan I. 2001. *Radical Enlightenment: Philosophy and the Making of Modernity, 1650–1750.* Oxford: Oxford University Press.

James, Frank A. L. J. 2005. "An 'Open Clash between Science and the Church'?: Wilberforce, Huxley and Hooker on Darwinism at the British Association, Oxford, 1860." In David Knight and Matthew D. Eddy, eds., *Science and Beliefs: From Natural Philosophy to Natural Science,* 171–194. Aldershot: Ashgate.

Jensen, J. Vernon. 1988. "Return to the Wilberforce-Huxley Debate." *British Journal of the History of Science,* 21: 161–179.

Johnson, Phillip E. 1991. *Darwin on Trial.* New York: Intervarsity Press and Regnery Gateway.

Jordanova, L. 1984. *Lamarck.* Oxford: Oxford University Press.

Kauffman, Stuart A. 1995. *At Home in the Universe: The Search for the Laws of Self-Organization and Complexity.* Oxford: Oxford University Press.

Kingsley, Charles. 1889. *The Water Babies.* New ed., London: Macmillan.

Kohn, David. 1980. "Theories to Work By: Rejected Theories, Reproduction, and Darwin's Path to Natural Selection." *Studies in the History of Biology* 4: 67–170.

————, ed. 1985. *The Darwinian Heritage: A Centennial Retrospect.* Princeton: Princeton University Press.

Kottler, Malcolm Jay. 1974. "Alfred Russel Wallace, the Origin of Man, and Spiritualism." *Isis* 65: 145–192.

————. 1985. "Charles Darwin and Alfred Russel Wallace: Two Decades of Debate over Natural Selection." In Kohn 1985, 367–432.

Kropotkin, Peter. 1902. *Mutual Aid: A Factor in Evolution*. London: reprinted with an introduction by Ashley Montagu, Boston: Extending Horizon Books.

Lack, David. 1947. *Darwin's Finches*. Cambridge: Cambridge University Press.

———. 1957. *Evolutionary Theory and Christian Belief: The Unresolved Conflict*. London: Methuen.

Lamarck, Jean-Baptiste Pierre Antoine de Monet, Chevalier de. 1914. *Zoological Philosophy*. Trans. Hugh Elliot. London: reprinted New York: Hafner, 1963.

Larson, Edward J. 1998. *Summer for the Gods: The Scopes Trial and America's Continuing Debate over Science and Religion*. New York: Basic Books/ Cambridge, MA: Harvard University Press.

Larson, James L. 1971. *Reason and Experience: The Representation of Natural Order in the Work of Carl von Linné*. Berkeley: University of California Press.

Laudan, Rachel. 1987. *From Mineralogy to Geology: The Foundations of a Science, 1650–1830*. Chicago: University of Chicago Press.

LeConte, Joseph. 1899. *Evolution: Its Nature, its Evidences and Its Relation to Religious Thought*. 2nd ed. New York: reprinted New York: Kraus, 1970.

Le Mahieu, D. L. 1976. *The Mind of William Paley: A Philosopher of His Age*. Lincoln, NE: University of Nebraska Press.

Lewis, Cherry. 2000. *The Dating Game: One Man's Search for the Age of the Earth*. Cambridge: Cambridge University Press.

Lightman, Bernard. 1987. *The Origins of Agnosticism: Victorian Unbelief and the Limits of Knowledge*. Baltimore: Johns Hopkins University Press.

Lindberg, David C., and Ronald L. Numbers, eds. 1986. *God and Nature: Historical Essays on the Encounter between Science and Religion*. Berkeley: University of California Press.

———. 2003. *When Science and Christianity Meet*. Chicago: University of Chicago Press.

Livingstone, David N. 1987. *Darwin's Forgotten Defenders: The Encounter between Evangelical Theology and Evolutionary Thought*. Edinburgh: Scottish Universities Press/Grand Rapids, MI: Eerdmans.

Livingstone, David N., D. G. Hart, and Mark A. Noll, eds. 1999. *Evangelicalism and Science in Historical Perspective*. New York: Oxford University Press.

Loewenberg, Bert James. 1969. *Darwin Comes to America: 1859–1900*. Philadelphia: Fortress Press.

Lopez, Michael. 1996. *Emerson and Power: Creative Antagonism in the Nineteenth Century*. DeKalb: Northern Illinois University Press.

Lucas, J. R. 1979. "Wilberforce and Huxley: A Legendary Encounter." *Historical Journal* 22: 313–330.

Lukas, Mary and Ellen Lukas. 1977. *Teilhard: The Man, the Priest, the Scientist.* New York: Doubleday.

Lurie, Edward. 1960. *Louis Agassiz: A Life in Science.* Chicago: University of Chicago Press.

Lyell, Charles. 1830–33. *Principles of Geology: Being an Attempt to Explain the Former Changes of the Earth's Surface by Reference to Causes Now in Operation.* 3 vols. London: reprinted with an introduction by M. J. S. Rudwick, Chicago: University of Chicago Press, 1990–91.

Lynch, John M., ed. 2000. *Vestiges and the Debate before Darwin.* 7 vols. Bristol: Thoemmes Press.

Lyons, Cherrie L. 1999. *Thomas Henry Huxley: The Evolution of a Scientist.* New York: Prometheus Books.

Machen, J. Gresham. 1937. *The Christian View of Man.* Reprinted London: The Banner and Truth Trust, 1967.

Mathews, Shailer. 1907. *The Church and the Changing Order.* New York: Macmillan.

———. 1924. *The Faith of Modernism.* New York: Macmillan.

———. 1936. *New Faith for Old: An Autobiography.* New York: Macmillan.

Mayr, Ernst, and William B. Provine, eds. 1980. *The Evolutionary Synthesis: Perspectives on the Unification of Biology.* Cambridge, MA: Harvard University Press. Rev. ed., 1998.

Mazumdar, Pauline. 1992. *Eugenics, Human Genetics and Human Failings: The Eugenics Society, Its Sources and Its Critics in Britain.* London: Routledge.

Mcgrath, Alistair. 2005. *Dawkin's God: Genes, Memes, and the Meaning of Life.* Oxford: Blackwell.

McMahon, Susan. 1999. "'In These Times of Giddiness and Distraction': The Shaping of John Ray and His Contemporaries, 1644–1662." In Nigel Cooper, ed., *John Ray and his Successors: The Clergyman as Biologist,* 80–94. Braintree: The John Ray Trust.

McNeil, Maureen. 1987. *Under the Banner of Science: Erasmus Darwin and His Age.* Manchester: Manchester University Press.

Medawar, P. B. 1982. *Pluto's Republic.* Oxford: Oxford University Press.

Messenger, Ernest. 1931. *Evolution and Theology: The Problem of Man's Origin.* London: Burns, Oates and Washbourne.

Millar, Ronald. 1972. *The Piltdown Men: A Case of Archaeological Fraud.* London: Victor Gollancz.

Miller, Hugh. 1841. *The Old Red Sandstone: Or New Walks in an Old Field*. Edinburgh. Boston, 1857 ed. reprinted New York: Arno Press, 1977.

———. 1850. *Footprints of the Creator Or the Asterolepis of Stromness*. 3rd ed. Edinburgh. Edinburgh, 1861 edition reprinted Farnborough: Gregg, 1971.

Miller, Kenneth R. 1999. *Finding Darwin's God: A Scientist's Search for Common Ground between God and Evolution*. New York: Harper Collins.

Millhauser, Milton. 1959. *Just before Darwin: Robert Chambers and* Vestiges. Middletown, CT: Wesleyan University Press.

Mivart, St. George Jackson. 1871. *The Genesis of Species*. London: Macmillan.

———. 1892. *Essays and Criticisms*. London: James R. Osgood, McIlvaine. 2 vols.

Monod, Jacques. 1971. *Chance and Necessity: An Essay on the Natural Philosophy of Modern Biology*. Trans. Austryn Wainhouse. New York: Vintage.

Montagu, Ashley, ed. 1982. *Evolution and Creation*. New York: Oxford University Press.

Monypenny, William F. and George E. Buckle. 1929. *The Life of Benjamin Disraeli*. 2 vols. Rev. ed. London: John Murray.

Moore, Aubrey. 1889. *Science and the Faith: Essays on Apologetic Subjects*. London: Kegan Paul, Trench.

Moore, James R. 1979. *The Post-Darwinian Controversies: A Study of the Protestant Struggle to Come to Terms with Darwin in Great Britain and America, 1870–1900*. New York: Cambridge University Press.

———. 1985a. "Herbert Spencer's Henchmen: The Evolution of Protestant Liberals in Late-Nineteenth-Century America." In Durant 1985, 76–100.

———. 1985b. "Evangelicals and Evolution: Henry Drummond, Herbert Spencer, and the Naturalization of the Spiritual World." *Scottish Journal of Theology* 38: 383–417.

———. 1989a. "Of Love and Death: Why Darwin 'Gave up Christianity.'" In Moore 1989b, 195–230.

———, ed. 1989b. *History, Humanity and Evolution: Essays for John C. Greene*. Cambridge: Cambridge University Press.

———. 1991. "Deconstructing Darwinism: The Politics of Evolution in the 1860s." *Journal of the History of Biology* 24: 353–408.

———. 1994. *The Darwin Legend*. Grand Rapids, MI: Baker Books.

Moore, John N. and Harold Slusher, eds. 1970. *Biology: A Search for Order in Complexity*. Grand Rapids, MI: Zonderevin.

Moorehead, Alan. 1969. *Darwin and the Beagle*. London: Hamish Hamilton.

Moran, Jeffrey P. 2003. "Reading Race into the Scopes Trial: African American Elites, Science and Fundamentalism." *Journal of American History,* 90: 891–911.

———. 2004. "The Scopes Trial and Southern Fundamentalism in Black and White America: Race, Region and Religion." *Journal of Southern History,* 70: 95–120.

Morgan, Conway Lloyd. 1923. *Emergent Evolution: The Gifford Lectures Delivered at the University of St. Andrews in the Year 1922.* London: Williams and Norgate.

Morris, H. M. 1974. *Scientific Creationism.* San Diego: Creation Life.

Morris, Simon Conway. 2003. *Life's Solution: Inevitable Humans in a Lonely Universe.* Cambridge: Cambridge University Press.

Muller, Hermann. 1959. "One Hundred Years without Darwin Are Enough." *The Humanist,* 19: 139–149.

Nelkin, Dorothy. 1977. *Science Textbook Controversies and the Politics of Equal Time.* Cambridge, MA: MIT Press.

———. 1983. *The Creation Controversy: Science or Scripture in Public Schools.* New York: Norton.

Numbers, Ronald L. 1977. *Creation by Natural Law: Laplace's Nebular Hypothesis in American Thought.* Seattle: University of Washington Press.

———. 1992. *The Creationists.* New York: Alfred A. Knopf.

———, ed. 1994–95. *Creationism in Twentieth-Century America.* 10 vols. New York: Goddard.

———. 1998. *Darwinism comes to America.* Cambridge, MA: Harvard University Press.

———. 2006. *The Creationists.* Rev. ed. Cambridge, MA: Harvard University Press.

Numbers, Ronald L. and John Stenhouse, eds. 1999. *Disseminating Darwinism: The Role of Place, Race, Religion and Gender.* Cambridge: Cambridge University Press.

O'Brien, Charles E. 1971. *Sir William Dawson: A Life in Science and Religion.* New York: Memoirs of the American Philosophical Society, no. 84.

Olby, Robert C. 1985. *The Origins of Mendelism.* 2nd ed. Chicago: University of Chicago Press.

Oldroyd, D. R. 1996. *Thinking about the Earth: A History of Geological Ideas.* London: Athlone.

O'Leary, Don. 2006. *Roman Catholicism and Modern Science: A History.* New York: Continuum.

Ospovat, Dov. 1981. *The Development of Darwin's Theory: Natural History, Natural Theology, and Natural Selection, 1838–59.* Cambridge: Cambridge University Press.

Outram, Dorinda. 1984. *Georges Cuvier: Vocation, Science, and Authority in Post-Revolutionary France.* Manchester: Manchester University Press.

———. 1995. *The Enlightenment.* Cambridge: Cambridge University Press.

Owen, Richard. 1849. *On the Nature of Limbs.* London: Van Voorst.

———. 1866–68. *The Anatomy of the Vertebrates.* 3 vols. London: reprinted New York: AMS Press, 1973.

Paley, William. 1802. *Natural Theology: Or Evidences of the Existence and Attributes of the Deity Collected from the Appearances of Nature.* London: reprinted Farnborough: Gregg, 1970.

Paradis, James G. 1978. *T. H. Huxley: Man's Place in Nature.* Lincoln, NE: University of Nebraska Press.

Paul, Harry W. 1974. "Religion and Darwinism: Varieties of Catholic Reaction." In Glick 1974, 403–436.

Pauly, Philip J. 1982. "Samuel Butler and His Darwinian Critics." *Victorian Studies* 25: 161–180.

Peacocke, A. R. 1980. *Creation and the World of Science.* Oxford: Oxford University Press.

———. 1986. *God and the New Biology.* London: Dent.

———. 1993. *Theology for a Scientific Age.* Enl. ed., London: SCM.

Peel, J. D. Y. 1971. *Herbert Spencer: The Evolution of a Sociologist.* London: Heinemann.

Pennock, Robert T. 2000. *Tower of Babel: The Evidence against the New Creationism.* Cambridge, MA: MIT Press.

———, ed. 2001. *Intelligent Design Creationism and its Critics: Philosophical, Theological and Scientific Perspectives.* Cambridge, MA: MIT Press.

Pfeifer, Edward J. 1965. "The Genesis of American Neo-Lamarckism." *Isis* 56:156–167.

———. 1974. "United States." In Glick 1974, 168–206.

Polkinghorne, John. 1988. *Science and Creation: The Search for Understanding.* London: SPCK.

———. 1989. *Science and Providence: God's Interaction with the World.* London: SPCK.

———. 1994. *Science and Christian Belief: Reflections of a Bottom-Up Thinker.* London: SPCK.

———. 1998a. *Science and Theology: An Introduction.* London: SPCK.

————. 1998b. *Belief in God in an Age of Science.* New Haven: Yale University Press.

————. 2000. *Faith, Science and Understanding.* London: SPCK.

Porte, Joel and Sandra Morris. 1999. *The Cambridge Companion to Ralph Waldo Emerson.* Cambridge: Cambridge University Press.

Porter, Roy. 1989. "Erasmus Darwin: Doctor of Evolution?" In J. R. Moore, ed., *History, Humanity and Evolution,* 39–70. Cambridge: Cambridge University Press.

————. 1990. *The Enlightenment.* Basingstoke: Macmillan. 2nd ed. 2000.

Powell, Baden. 1855. *Essays on the Spirit of the Inductive Philosophy, the Unity of Worlds and the Philosophy of Creation.* London: reprinted Farnborough: Gregg International, 1969.

Provine, William B. 1971. *The Origins of Theoretical Population Genetics.* Chicago: University of Chicago Press.

Raven, Charles E. 1927. *The Creator Spirit: A Survey of Christian Doctrine in the Light of Biology, Psychology and Mysticism.* London: Martin Hopkinson.

————. 1942. *John Ray, Naturalist: His Life and Work.* Cambridge: Cambridge University Press.

————. 1953. *Natural Religion and Christian Theology.* Cambridge: Cambridge University Press.

————. 1962. *Teilhard de Chardin: Scientist and Seer.* London: Collins.

Ray, John. 1691. *The Wisdom of God as Manifested in the Works of Creation.* London. 1717 ed. reprinted New York: Arno Press, 1977.

————. 1692. *Miscellaneous Discourses concerning the Changes of the World.* London: reprinted Hildesheim: Georg Olms, 1968.

————. 1713. *Three Physico-Theological Discourses.* 3rd ed. London: reprinted New York: Arno Press, 1977.

Reader, J. 1981. *Missing Links: The Hunt for Earliest Man.* London: Collins.

Regal, Brian. 2002. *Henry Fairfield Osborn: Race, and the Search for the Origins of Man.* Aldershot: Ashgate.

Richards, Robert J. 1987. *Darwin and the Emergence of Evolutionary Theories of Mind and Behavior.* Chicago: University of Chicago Press.

Roberts, Jon H. 1988. *Darwin and the Divine in America: Protestant Intellectuals and Organic Evolution, 1859–1900.* Madison: University of Wisconsin Press.

Roger, Jacques. 1997. *Buffon: A Life in Natural History.* Trans. Sarah L. Bonnefoi, ed. L. Pierce Williams. Ithaca: Cornell University Press.

————. 1998. *The Life Sciences in Eighteenth-Century French Thought.*

Trans. Robert Ellrich, ed. Keith R. Benson. Stanford: Stanford University Press.

Rolston, Homes, III. 1999. *Genes, Genesis and God: Values and their Origin in Natural and Human History.* Cambridge: Cambridge University Press.

Rudwick, Martin J. S. 1972. *The Meaning of Fossils: Episodes in the History of Paleontology.* 2nd ed. New York: Science History Publications, 1976.

———. 1985. *The Great Devonian Controversy: The Shaping of Scientific Knowledge among Gentlemenly Specialists.* Chicago: University of Chicago Press.

———. 2005. *Bursting the Limits of Time: The Reconstruction of Geohistory in the Age of Revolution.* Chicago: University of Chicago Press.

Rupke, Nicolaas A. 1983. *The Great Chain of History: William Buckland and the English School of Geology (1814–1849).* Oxford: Oxford University Press.

———. 1994. *Richard Owen: Victorian Naturalist.* New Haven: Yale University Press.

Ruse, Michael. 1979. *The Darwinian Revolution: Science Red in Tooth and Claw.* Chicago: University of Chicago Press. 2nd ed. 1999.

———. 1996. *Monad to Man: The Concept of Progress in Evolutionary Biology.* Cambridge, MA: Harvard University Press.

———. 2001. *Can a Darwinian Be a Christian? The Relationship between Science and Religion.* Cambridge, MA: Harvard University Press.

———. 2003. *Darwin and Design: Does Evolution Have a Purpose?* Cambridge, MA: Harvard University Press.

———. 2005. *The Evolution-Creation Struggle.* Cambridge, MA: Harvard University Press.

Russett, Cynthia Eagle. 1976. *Darwin in America: The Intellectual Response.* San Francisco: W. H. Freeman.

Scopes, John Thomas. 1967. *Center of the Storm.* New York: Holt, Rinehart and Winston.

Scott, Clifford H. 1976. *Lester Frank Ward.* Boston: Twayne Publishers.

Scott, Eugenie C. 2004. *Evolution vs. Creationism: An Introduction.* Berkeley: National Center for Science Education.

Secord, James A. 2000. *Victorian Sensation: The Extraordinary Publication, Reception and Secret Authorship of* Vestiges of the Natural History of Creation. Chicago: University of Chicago Press.

Settle, M. L. 1972. *The Scopes Trial.* New York: Franklin Watts.

Shanks, Niall. 2004. *God, The Devil, and Darwin: A Critique of Intelligent Design Theory.* New York: Oxford University Press.

Shaw, George Bernard. 1921. *Back to Methuselah: A Metabiological Pentateuch.* London: Constable.

Sheets-Pyenson, Susan. 1996. *John William Dawson: Faith, Hope, Science.* Montreal: McGill-Queen's University Press.

Simpson, George Gaylord. 1963. *This View of Life; the World of an Evolutionist.* New York: Harcourt, Brace and World.

Sloan, Phillip R. 1985. "Darwin's Invertebrate Program, 1826–1836." In Kohn 1985, 71–120.

———. 1986. "Darwin, Vital Matter, and the Transformism of Species." *Journal of the History of Biology,* 19: 369–445.

Smith, Crosbie W. and M. Norton Wise. 1989. *Energy and Empire: A Biographical Study of Lord Kelvin.* Cambridge: Cambridge University Press.

Spencer, Frank. 1990. *Piltdown: A Scientific Forgery.* London: Oxford University Press.

Spencer, Herbert. 1851. *Social Statics: Or the Conditions Essential to Human Happiness Specified, and One of Them Adopted.* London: reprinted Farnborough: Gregg International.

———. 1855. *Principles of Psychology.* London: reprinted Farnborough: Gregg International. 1881 ed. reprinted (2 vols.) Boston: Longwood Press.

———. 1862. *First Principles of a New Philosophy.* London.

———. 1864. *Principles of Biology.* 2 vols. London.

———. 1883. *Essays Scientific, Political and Speculative.* 3 vols. London.

Stepan, Nancy. 1982. *The Idea of Race in Science: Great Britain, 1800–1960.* London: Macmillan.

Stephens, Lester G. 1982. *Joseph LeConte: Gentle Prophet of Evolution.* Baton Rouge: Louisiana State University Press.

Stephenson, Alan M. G. 1984. *The Rise and Decline of English Modernism.* London: SPCK.

Sulloway, Frank J. 1982. "Darwin and His Finches: The Evolution of a Legend." *J. Hist. Biology* 15: 1–54.

Swetlitz, Marc. 1995. "Julian Huxley and the End of Evolution." *Journal of the History of Biology* 28: 181–217.

Teilhard de Chardin, Pierre. 1959. *The Phenomenon of Man.* Introd. Julian Huxley. London: Collins.

Turner, Frank Miller. 1974. *Between Science and Religion: The Reaction to*

Scientific Naturalism in Late Victorian England. New Haven: Yale University Press.

————. 1978. "The Victorian Conflict between Science and Religion: A Professional Dimension." *Isis* 69: 356–376.

Turrill, William Bertram. 1963. *Joseph Dalton Hooker: Botanist, Explorer and Administrator.* London: Scientific Book Guild.

Van Riper, A. Bowdoin. 1993. *Men among the Mammoths: Victorian Science and the Discovery of Human Prehistory.* Chicago: University of Chicago Press.

Vorzimmer, Peter J. 1970. *Charles Darwin, the Years of Controversy: The* Origin of Species *and Its Critics, 1859–82.* Philadelphia: Temple University Press.

Walker, A. and P. Shipman. 1996. *The Wisdom of Bones: In Search of Human Origins.* New York: Knopf.

Wallace, Alfred Russel. 1870. *Contributions to the Theory of Natural Selection.* London: reprinted New York: AMS Press, 1973.

————. 1889. *Darwinism: An Exposition of the Theory of Natural Selection.* London: reprinted New York, AMS Press, 1975.

————. 1891. *Natural Selection and Tropical Nature.* London: reprinted Farnborough: Gregg International, 1969.

Wallace, David Rains. 1999. *The Bonehunters' Revenge: Dinosaurs, Greed, and the Greatest Scientific Fraud of the Gilded Age.* Boston: Houghton Mifflin.

Ward, Keith. 1996. *God, Chance and Necessity.* Oxford: One World.

Warfield, B. B. 2000. *Evolution, Science and Scripture: Selected Writings.* Ed. Mark A. Noll and David N. Livingstone. Grand Rapids, MI: Baker Books.

Waters, C. Kenneth and Albert van Helden, eds. 1992. *Julian Huxley: Biologist and Statesman of Science.* Houston: Rice University Press.

Weiner, J. S. 1955. *The Piltdown Hoax.* Oxford: Oxford University Press.

Whitcomb, J. C., and H. M. Morris. 1961. *The Genesis Flood.* Nutley, NJ: Presbyterian and Reformed Publishing Co.

White, Andrew Dickson. 1896. *A History of the Warfare of Science with Theology.* 2 vols. Reprinted New York: Dover, 1969.

Whitehead, Alfred North. 1929. *Process and Reality: An Essay in Cosmology.* Cambridge: Cambridge University Press.

Willey, Basil. 1960. *Darwin and Butler: Two Versions of Evolution.* London: Chatto and Windus.

Wilson, Edward O. 1975. *Sociobiology: The New Synthesis.* Cambridge, MA: Harvard University Press.

————. 1978. *On Human Nature.* Cambridge, MA: Harvard University Press.

————. 1992. *The Diversity of Life.* Cambridge, MA: Harvard University Press, reprinted Harmondsworth: Penguin.

Wilson, Leonard G. 1972. *Charles Lyell: The Years to 1841: The Revolution in Geology.* New Haven: Yale University Press.

Young, Robert M. 1985. *Darwin's Metaphor: Nature's Place in Victorian Culture.* Cambridge: Cambridge University Press.

Zahm, John. 1896. *Evolution and Dogma.* Chicago: McBride and Co.